T0186028

An Introduction to the Theory of Superfluidity

An Introduction to the Theory of Superfluidity

I. M. Khalatnikov
Moscow Physiotechnical Institute

Translated from the Russian edition by
Pierre C. Hohenberg

This book was originally published as part of the Frontiers in Physics Series, edited by David Pines.

Advanced Book Program

CRC Press
Taylor & Francis Group
Boca Raton London New York

CRC Press is an imprint of the
Taylor & Francis Group, an **informa** business

First published 1965 by Westview Press

Published 2018 by CRC Press
Taylor & Francis Group
6000 Broken Sound Parkway NW, Suite 300
Boca Raton, FL 33487-2742

CRC Press is an imprint of the Taylor & Francis Group, an informa business

Copyright © 1965, 1989, 2000 by I. M. Khalatnikov

No claim to original U.S. Government works

This book contains information obtained from authentic and highly regarded sources. Reasonable efforts have been made to publish reliable data and information, but the author and publisher cannot assume responsibility for the validity of all materials or the consequences of their use. The authors and publishers have attempted to trace the copyright holders of all material reproduced in this publication and apologize to copyright holders if permission to publish in this form has not been obtained. If any copyright material has not been acknowledged please write and let us know so we may rectify in any future reprint.

Except as permitted under U.S. Copyright Law, no part of this book may be reprinted, reproduced, transmitted, or utilized in any form by any electronic, mechanical, or other means, now known or hereafter invented, including photocopying, microfilming, and recording, or in any information storage or retrieval system, without written permission from the publishers.

Trademark Notice: Product or corporate names may be trademarks or registered trademarks, and are used only for identification and explanation without intent to infringe.

Visit the Taylor & Francis Web site at
http://www.taylorandfrancis.com

and the CRC Press Web site at
http://www.crcpress.com

A CIP catalog record for this book is available from the Library of Congress.

ISBN 13: 978-0-7382-0300-3 (pbk)

Advanced Book Classics

David Pines, Series Editor

Anderson, P.W., *Basic Notions of Condensed Matter Physics*
Bethe H. and Jackiw, R., *Intermediate Quantum Mechanics, Third Edition*
Cowan, G. and Pines, D., *Complexity: Metaphors, Models, and Reality*
de Gennes, P.G., *Superconductivity of Metals and Alloys*
d'Espagnat, B., *Conceptual Foundations of Quantum Mechanics, Second Edition*
Feynman, R., *Photon-Hadron Interactions*
Feynman, R., *Quantum Electrodynamics*
Feynman, R., *Statistical Mechanics*
Feynman, R., *The Theory of Fundamental Processes*
Gell-Mann, M. and Ne'eman, Y., *The Eightfold Way*
Khalatnikov, I. M. *An Introduction to the Theory of Superfluidity*
Ma, S-K., *Modern Theory of Critical Phenomena*
Migdal, A. B., *Qualitative Methods in Quantum Theory*
Negele, J. W. and Orland, H., *Quantum Many-Particle Systems*
Nozières, P., *Theory of Interacting Fermi Systems*
Nozières, P. and Pines, D., *The Theory of Quantum Liquids*
Parisi, G., *Statistical Field Theory*
Pines, D., *Elementary Excitations in Solids*
Pines, D., *The Many-Body Problem*
Quigg, C., *Gauge Theories of the Strong, Weak, and Electromagnetic Interactions*
Schrieffer, J.R., *Theory of Superconductivity, Revised*
Schwinger, J., *Particles, Sources, and Fields, Volume I*
Schwinger, J., *Particles, Sources, and Fields, Volume II*
Schwinger, J., *Particles, Sources, and Fields, Volume III*
Schwinger, J., *Quantum Kinematics and Dynamics*
Wyld, H.W., *Mathematical Methods for Physics*

Special Preface

Almost 25 years have elapsed since publication of the first version of this book. As the author of the book, I believe that the book, written as an introduction to the theory of the field of physics so advanced for this period, has not lost its importance. No essential changes have occurred in the fundamentals of the superfluidity theory based on Landau's ideas.

The book consists of 4 parts. Part 1 covers main properties of the excitation spectrum in superfluid ^{4}He and the thermodynamics determined by the spectrum. The material presented in Sections 1-6 of Part 1 needs no additional comments. Section 7 is devoted to the description of the interaction of elementary excitations. In reading this section one should bear in mind that our notions of the long wavelength phonon part of the spectrum have undergone certain alterations. Experiment has revealed that in the initial part of the phonon spectrum the sign of dispersion is such that conservation laws permit decay of one phonon into two. Therefore, although this does not affect qualitative results of the calculations of the kinetic effects, described in Part 3, quantitative results should be insignificantly modified with processes of phonon decay taken into account.

Part 2, "Hydrodynamics," has remained, as it should, invariant and no statement has lost its importance.

When working with Part 3, one should remember the remark concerning the phonon decay effect[1].

A brief Part 4 is devoted to solutions of impurities in superfluid ^4He. As an introduction to this field it retains its validity.

The book was intended for young researchers involved in theory and experiment of low temperature physics.

Professor I.M. Khalatnikov
1989

[1] See I.M. Khalatnikov "Phenomenological Theory of Superfluid ^4He "in the book "The Physics of Liquid and Solid Helium," Part 1, ed. by K.H. Bennemann & J.B. Ketterson, John Wiley & Sons, New York

Editor's Foreword

Perseus Publishing's *Frontiers in Physics* series has, since 1961, made it possible for leading physicists to communicate in coherent fashion their views of recent developments in the most exciting and active fields of physics—without having to devote the time and energy required to prepare a formal review or monograph. Indeed, throughout its nearly forty year existence, the series has emphasized informality in both style and content, as well as pedagogical clarity. Over time, it was expected that these informal accounts would be replaced by more formal counterparts—textbooks or monographs—as the cutting-edge topics they treated gradually became integrated into the body of physics knowledge and reader interest dwindled. However, this has not proven to be the case for a number of the volumes in the series: Many works have remained in print on an on-demand basis, while others have such intrinsic value that the physics community has urged us to extend their life span.

The *Advanced Book Classics* series has been designed to meet this demand. It will keep in print those volumes in *Frontiers in Physics* that continue to provide a unique account of a topic of lasting interest. And through a sizable printing, these classics will be made available at a comparatively modest cost to the reader.

The lectures contained in *An Introduction to the Theory of Superfluidity* provide an unusually lucid and complete account of the essentials of one of the most fascinating phenomena in physics, the ability of a liquid to flow without resistance at low temperatures. Written by Isaac M. Khalatnikov, a leading researcher in superfluidity, who was a close collaborator with the great Russian theoretical physicist, Lev D. Landau, and the first Director of the world-famous Institute that bears Landau's name, these lectures have served to introduce countless students of low temperature physics to the physics of superfluid liquid Helium 4. I am pleased that their publication in *Advanced Book Classics* will make the lectures readily available to future generations of graduate students and experienced researchers.

David Pines
Cambridge, England
May, 2000

Vita

Isaac Markovich Khalatnikov

Isaac Markovich Khalatnikov is a Professor of Theoretical Physics at the Moscow Physiotechnical Institute. A graduate of Dniepropetrovsk University, Dr. Khalatnikov worked in the theoretical section of the Institute of Physical Problems at the U.S.S.R. Academy of Sciences under the direction of L.D. Landau. Khalatnikov and Landau together founded the theory of quantum liquids. Khalatnikov has worked as director of the L.D. Landau Institute of Sciences and has been appointed as an academician at the academy. Dr. Khalatnikov is a U.S.S.R. State Prize Laureate and is a past winner of the Landau Prize of the U.S.S.R. Academy of Sciences. His main research interests are in the fields of superconductivity and superfluidity theory, quantum field theory and cosmology.

Preface

The present book is an exposition of the modern theory of superfluidity, a phenomenon which occupies a distinctive place in contemporary physics. This is first of all due to the fact that superfluidity is a macroscopic manifestation of quantum laws. The phenomenon of superfluidity was at first thought to be rather exotic, and restricted to liquid helium. It is now known that, in one form or another, superfluidity is found in all macroscopic bodies, wherever quantum laws are applicable.

Intensive interest in this phenomenon has led to progress in solid state physics as a whole. The ideas and methods of the theory of superfluidity have turned out to bear fruit in many branches of physics, quite far removed from solid state physics, such as the theory of nuclear structure.

This book is intended for research workers and graduate students, and may serve as an introduction to this most interesting field of contemporary physics. It is assumed that the reader is familiar only with the fundamentals of quantum mechanics and statistical physics.

The author wishes to express his deep gratitude to Professor David Pines, who first suggested writing this book. The author is likewise grateful to Dr. Pierre Hohenberg for his considerable labor in translating and editing the manuscript.

I.M.K.
Moscow
April 1965

Translator's Preface

The task of translating this volume would have been much less rewarding for me had I not had the pleasure of meeting the author, and discussing certain questions with him in person. I would like to take this opportunity to thank Professor Khalatnikov for the unforgettable experience of working for a year within the theoretical group at the Institute for Physical Problems.

P.C.H.

Contents

An Introduction to the Theory of Superfluidity

PART I

ELEMENTARY EXCITATIONS

1

THE ENERGY SPECTRUM OF A QUANTUM LIQUID AND SUPERFLUIDITY[1,2,3]

At a temperature of 2.18°K, liquid helium undergoes a second-order phase transition. Below the λ-point, liquid helium (helium II) has a number of unusual properties, of which the most remarkable is superfluidity, discovered by P. Kapitza. This is the ability of liquid helium to flow without friction through narrow capillaries. It is easy to convince oneself that at temperatures of the order of 1 or 2°K, the de Broglie wavelength of helium atoms is comparable to the inter-atomic distance. It follows that helium II has quantum properties; it is therefore not a classical liquid, but a quantum liquid. As is well known, there are two stable isotopes of helium, He^3 and He^4, of mass 3 and 4, respectively, in atomic units. The liquid which exhibits superfluidity is the one formed from atoms of He^4—that is, from particles obeying Bose statistics. He^3 atoms also form a quantum liquid which, however, does not exhibit superfluidity in the above-mentioned temperature region. A quantum liquid made up of Fermi particles is usually called a Fermi liquid. We may therefore say that only liquids made up of Bose particles possess the property of superfluidity.

In recent years, however, it has become clear that in a Fermi liquid consisting of atoms of He^3, at sufficiently low temperatures (probably around 0.001°K), pairing should occur—that is, the formation of particles of Bose type. This should lead to the occurrence of superfluidity. In this manner one gains the impression that the property of superfluidity is in one form or another a feature of all quantum liquids. The list of macroscopic objects that are quantum liquids is limited to the two above-mentioned liquids made up of isotopes of helium.† However, we encounter the properties of superfluidity in other systems. For example, electrons in metals can at low temperature form pairs—that is, Bose particles; this leads to the superfluidity of the electron "liquid," which appears in the form of superconductivity, since the liquid is charged. We thus see that the property of superfluidity occurs in a number of quantum systems at low temperatures and is not as exotic as was thought earlier. The link between the phenomenon of superfluidity and the quantum properties of a system, as well as the resulting theory of the phenomenon, were first established by L. Landau.[1,2,3]

According to classical mechanics, at zero temperature all atoms should be at rest and their potential energy should be minimum. Consequently, at low temperatures, they can only undergo small oscillations about certain equilibrium positions. This means that at low temperatures all bodies should solidify and form a crystalline lattice. Liquid helium is the only system where quantum effects appear before the liquid has solidified. This is due to the relatively weak interaction between helium atoms. In all other media the interaction between atoms is sufficiently strong so that the body solidifies before quantum effects have appeared. At zero temperature the system of atoms forming a solid body is in the lowest or ground state of energy; at higher temperatures the system makes transitions to excited states. The atoms undergo oscillations around their equilibrium positions. The energy of the crystal is the sum of the energies of certain quantum oscillators, each one of these being in one or another of its excited states. One may also look upon an oscillator in its n-th excited state as a collection of n quanta. These quanta, or phonons, correspond to sound waves in the same way as photons correspond to light waves. In this manner the state of the system is characterized by a set of long-wavelength sound quanta or phonons. These have an energy proportional to their momentum (or more precisely their "quasi-momentum"). With the use of the phonon representation one can explain all the low-temperature properties of solids.

The situation described above is not unique in quantum mechanics. Any system of particles with arbitrary interactions can, in its weakly excited states, be looked upon as a set of distinct elementary excitations. Each elementary excitation behaves like a quasi-particle, capable of motion throughout the body. A quasi-particle has a definite energy and momentum. The function which characterizes the dependence of the energy on the momentum is called the energy spectrum of the body.

---

†The rapidly decaying isotope He^6 has not been studied extensively; however, a liquid consisting of atoms of He^6 would undoubtedly possess the property of superfluidity.

Let us denote by ϵ the energy of an elementary excitation in liquid helium as a function of its momentum p. The form of the energy spectrum for small values of the momentum p is easily determined. Small momenta correspond to long-wavelength excitations which in a liquid are obviously just longitudinal sound waves. The corresponding elementary excitations are therefore sound quanta or phonons, whose energy is a linear function of the momentum

$$\epsilon = cp \qquad (1\text{-}1)$$

where c is the velocity of sound. As the momentum increases the curve $\epsilon(p)$ departs from a straight line. Its subsequent behavior, however, cannot be obtained from general considerations. In order to explain the experimental values obtained for the thermodynamic functions of liquid helium, L. Landau proposed the energy spectrum shown in Fig. 1. It turned out that the phonons alone were not sufficient to explain the temperature dependence and the absolute values of such thermodynamic quantities as the specific heat, for instance. It is easy to see that elementary excitations with energies close to the minimum on the curve of Fig. 1 will give a contribution to all thermodynamic quantities, which competes with the contribution of the phonons. The corresponding excitations were called rotons and their energy could be represented near the minimum in the form

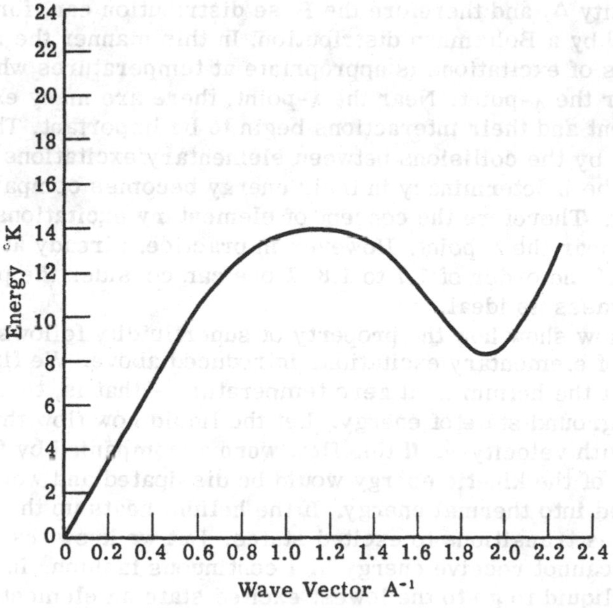

Figure 1. The energy spectrum of liquid helium II.

$$\epsilon = \Delta + \frac{(p - p_0)^2}{2\mu} \tag{1-2}$$

Here p_0 is the value of the momentum at which the function ϵ has a minimum equal to Δ. The exact values of the parameters characterizing the energy spectrum of liquid helium were found by neutron scattering experiments. Monochromatic neutrons emit or absorb elementary excitations in helium. By measuring the energies of neutrons scattered at given angles one can determine the whole spectrum of elementary excitations. In this manner the following values of the spectrum parameters were obtained

$$\Delta/k = 8.6^\circ K \qquad p_0/\hbar = 1.91\ A^{-1} \qquad \mu = 0.16\ m_{He}$$

The quantity μ which has the dimensions of mass is usually called the effective mass of the roton.

The concept of elementary excitations can be used if few of these are present so that their interaction energy is small compared to their own energy. In this case the gas of elementary excitations can be looked upon as an ideal gas. Since upon excitation of the liquid, phonons and rotons can appear one at a time, it is obvious that they should have integer spin and, therefore, obey Bose statistics. Thus at equilibrium the phonon and roton gases are described by the equilibrium functions of Bose statistics. The roton energies contain the large quantity Δ, and therefore the Bose distribution can, for rotons, be replaced by a Boltzmann distribution. In this manner the model of an ideal gas of excitations is appropriate at temperatures which are not too near the λ-point. Near the λ-point, there are many excitations present and their interactions begin to be important. The lifetime determined by the collisions between elementary excitations becomes small and the indeterminacy in their energy becomes comparable with this energy. Therefore the concept of elementary excitations is not applicable near the λ-point. However in practice, already at temperatures of the order of 1.7 to 1.8°K one can consider the phonon and roton gases as ideal.

Let us now show how the property of superfluidity follows from the notion of elementary excitations introduced above. We first assume that the helium is at zero temperature—that is, that it occupies the ground state of energy. Let the liquid now flow through a capillary with velocity $\mathbf{v}$. If this flow were accompanied by friction, then a part of the kinetic energy would be dissipated and would be transformed into thermal energy. If the helium heats up this means that it makes transitions to excited states. But we know that a quantum liquid cannot receive energy in a continuous fashion. In order for such a liquid to go to the lowest excited state an elementary excitation must be created. Let the energy of such an excitation be

$\epsilon(p)$ and the corresponding momentum be p, in the frame moving with the liquid. Then in the fixed frame of reference the energy of the system has changed by the amount

$$\epsilon(p) + \mathbf{p}\cdot\mathbf{v} \tag{1-3}$$

The transition to an excited state will be energetically favorable if the condition

$$\epsilon(p) + \mathbf{p}\cdot\mathbf{v} < 0 \tag{1-4}$$

is fulfilled. Obviously, the most favorable situation is one in which the momentum of the created excitation is directed oppositely to the velocity. In that case it follows from Eq. 1-4 that

$$v > \frac{\epsilon(p)}{p} \tag{1-5}$$

Obviously an excitation can appear in liquid helium if condition Eq. 1-5 is fulfilled at least at that point in the spectrum where the ratio $\epsilon(p)/p$ has its minimum value. We thus obtain the necessary condition for the creation of an excitation

$$v > \min \frac{\epsilon(p)}{p} \tag{1-6}$$

Superfluidity will occur if

$$v_{cr} = \min \frac{\epsilon(p)}{p} \neq 0 \tag{1-7}$$

In that case, for values of the velocity of motion v of the helium which are less than v_{cr}, the creation of an excitation will be energetically unfavorable and the liquid will flow without dissipation of energy—that is, without friction. Frictionless flow will not slow down, that is it will be superfluid. On the other hand, at velocities greater than v_{cr}, the motion of the helium will cause the creation of excitations and consequently dissipation of energy. As can be seen from Fig. 1, the spectrum of elementary excitations of helium II satisfies condition Eq. 1-7. The minimum of the ratio ϵ/p can be found by the extremal condition,

$$\frac{d}{dp}\frac{\epsilon}{p} = \frac{1}{p}\frac{d\epsilon}{dp} - \frac{\epsilon}{p^2} = 0$$

to occur at the point where

$$\frac{\epsilon}{p} = \frac{d\epsilon}{dp} \tag{1-8}$$

that is, at the point where a straight line going through the origin of coordinates is tangent to the curve $\epsilon(p)$. For the spectrum of liquid helium this point lies near the minimum of the curve $\epsilon(p)$ and is numerically equal to about 60 m/sec. This value is several orders of magnitude larger than the actually observed critical velocity in helium II. This appears to be explainable by the fact that, as we shall see below, phonons and rotons are not the only excitations that can appear in superfluid helium; there can exist excitations of another type — so-called quantum vortices. Although these excitations have too little statistical weight to play a role in the thermodynamics, they are easily created and consequently are important in determining hydrodynamic properties such as critical velocities.† We may note that in deriving criterion Eq. 1-6 we nowhere used the fact that there were no gas-like excitations already present in the liquid. The presence of such excitations would not make the derivation invalid and therefore criterion (1-6) will also be true at finite temperatures. However, the presence of excitations already in the liquid at finite temperatures, introduces special features into the flow of helium II through capillaries. The excitations present in the liquid will be reflected against the walls and will transfer part of their momentum to the walls. Because of this, that part of the liquid which is carried along by the motion of the excitations will behave like a normal viscous liquid and will be slowed down by friction with the walls. Therefore at $T = 0$ the whole liquid flows through the capillary without friction, but for $T \neq 0$ only part of the liquid does. In this manner we obtain the remarkable situation, that in a superfluid two independent motions can take place simultaneously. One part, which is carried along by the excitations, behaves like a normal liquid and the rest, the superfluid, does not experience friction, as long as its velocity is less than some critical value, it moves independently of the normal part. There are thus two simultaneous motions possible in helium II — superfluid motion with velocity $\mathbf{v}_s$ and normal motion with velocity $\mathbf{v}_n$. To each one of these motions there corresponds a different effective mass. The sum of the normal and superfluid

---

†The critical velocity connected with the creation of quantum vortices depends on the diameter of the capillary according to the relation

$$v_{cr} \cdot d \sim h$$

This is why one must use narrow capillaries in order to observe superfluidity. In thick pipes the value of v_{cr} is too small for superfluidity to be observable.

masses is the total mass of the liquid. The two motions are independent of one another (at least in the region of small velocities $\mathbf{v}_s$ and $\mathbf{v}_n$, smaller than certain critical values, at which momentum transfer between the two parts becomes possible). The momentum per unit volume $\mathbf{j}$ of helium II is thus divided into two parts

$$\mathbf{j} = \rho_s \mathbf{v}_s + \rho_n \mathbf{v}_n \tag{1-9}$$

The coefficient ρ_n is called the normal density and ρ_s the superfluid density. Their sum is the total density of the liquid

$$\rho = \rho_n + \rho_s \tag{1-10}$$

The ratio ρ_n/ρ is unity at the λ-point; as the temperature decreases it decreases and it is zero at $T = 0$.

Let us suppose that in helium there are two simultaneous translational motions with velocities $\mathbf{v}_n$ and $\mathbf{v}_s$. The energy of an elementary excitation in the rest frame $E(p)$, can be expressed in terms of its energy $\epsilon(p)$ in the frame in which the superfluid is at rest by the relation

$$E(p) = \epsilon(p) + \mathbf{p} \cdot \mathbf{v}_s \tag{1-11}$$

The normal motion of the liquid is associated with the translational motion of the gas of excitations, occurring with the velocity $\mathbf{v}_n$. The distribution function n of the gas of elementary excitations depends on the energy of relative motion

$$E' = E - \mathbf{p} \cdot \mathbf{v}_n = \epsilon(p) + \mathbf{p} \cdot \mathbf{v}_s - \mathbf{p} \cdot \mathbf{v}_n \tag{1-12}$$

In moving helium II the energy distribution of phonons is therefore determined by the Planck function

$$n = \left[\exp\left(\frac{\epsilon + \mathbf{p} \cdot \mathbf{v}_s - \mathbf{p} \cdot \mathbf{v}_n}{kT}\right) - 1\right]^{-1} \tag{1-13}$$

where the function ϵ is given in Eq. 1-1.

The energy distribution of rotons is determined by the Boltzmann function

$$n = \exp\left(-\frac{\epsilon + \mathbf{p} \cdot \mathbf{v}_s - \mathbf{p} \cdot \mathbf{v}_n}{kT}\right) \tag{1-14}$$

where ϵ is given by Eq. 1-2.

2

THE THERMODYNAMIC FUNCTIONS OF HELIUM II

At temperatures not too near the λ-point the phonon and roton densities are small and, as was already mentioned, one may consider the gases of excitations as ideal gases. In that case all thermodynamic functions consist of two parts, one due to phonons and the other due to rotons.

In order to calculate the thermodynamic functions we use the distribution functions for phonons and rotons, Eqs. 1-13 and 1-14. As a first approximation we may neglect the dependence of the distribution functions on the relative velocity $\mathbf{v}_n - \mathbf{v}_s$. This dependence only begins to be important for a range of values of $\mathbf{v}_n - \mathbf{v}_s$ at which, under usual conditions, the destruction of superfluidity begins to set in. Only when sound waves of large amplitude are propagating through helium does the relative velocity $\mathbf{v}_n - \mathbf{v}_s$ attain large values. In treating that problem it is essential to take into account the quadratic terms in $\mathbf{v}_n - \mathbf{v}_s$ in the thermodynamic functions.

Let us first calculate the thermodynamic functions for helium II at rest. These can be obtained by using the general formulas of Bose statistics. As we stated previously the phonons obey Bose statistics and the distribution function of the rotons is independent of statistics because of the presence of the large constant term $\Delta \gg kT$ in the roton energy.

The free energy of a Bose gas is given by

$$F = -kT \int \ln(1 + n)\, d\tau_p \qquad (2\text{-}1)$$

where $d\tau_p = p^2\, dp\, do/(2\pi\hbar)^3$ is the volume element in p-space and do the element of solid angle. After integrating (2-1) by parts we obtain

$$F = -\frac{1}{3} \int np \frac{\partial \epsilon}{\partial p} d\tau_p \tag{2-2}$$

which allows one to calculate the free energy of the excitation gas. The entropy of the excitations is found by differentiating the free energy with respect to temperature

$$S = -\frac{\partial F}{\partial T} = -\frac{1}{3kT^2} \int n' \epsilon p \frac{\partial \epsilon}{\partial p} d\tau_p \tag{2-3}$$

(n' is the derivative of the distribution function with respect to its argument).

1 THE FREE ENERGY, ENTROPY, AND SPECIFIC HEAT OF THE PHONON GAS

If we carry out the integral in Eq. 2-2 using the distribution function Eq. 1-13 we find the free energy of the phonon gas ($\epsilon = cp$)

$$F_{ph} = -\frac{1}{3} \int (e^{\epsilon/kT} - 1)^{-1}\, p \frac{\partial \epsilon}{\partial p} d\tau_p = -\tfrac{1}{3} E_{ph} \tag{2-4}$$

The energy per unit volume of the phonons is equal to

$$E_{ph} = \frac{4\pi^5}{15} \left(\frac{kT}{2\pi\hbar c} \right)^3 kT \approx \frac{\pi^4}{36} kT \mathfrak{N}_{ph} \tag{2-5}$$

where $\mathfrak{N}_{ph} \approx 2.4 \times 4\pi (kT/2\pi\hbar c)^3$ is the number of phonons per unit volume in helium II. The entropy per unit volume can be found either by the general formula, Eq. 2-3, or directly by differentiating the relation obtained for the free energy. We thus find

$$S_{ph} = -\frac{\partial F_{ph}}{\partial T} = \frac{16\pi^5}{45} k \left(\frac{kT}{2\pi\hbar c} \right)^3 \tag{2-6}$$

We may furthermore calculate the specific heat of the phonon gas

$$C_{ph} = T \frac{\partial S_{ph}}{\partial T} = \frac{16\pi^5}{15} k \left(\frac{kT}{2\pi\hbar} \right)^3 \tag{2-7}$$

2 THE FREE ENERGY, ENTROPY, AND SPECIFIC HEAT OF THE ROTON GAS

The free energy per unit volume for the roton gas can be found by the general formula, Eq. 2-2, using the distribution function Eq. 1-14. In carrying out the integration it is necessary to take into account the fact that the roton momenta are in magnitude close to p_0. In this manner we find

$$F_r = -kT\mathfrak{N}_r \tag{2-8}$$

where $\mathfrak{N}_r$ is the number of rotons per unit volume of helium

$$\mathfrak{N}_r = \int n\, d\tau_p = \frac{2p_0^2(\mu kT)^{1/2}\, e^{-\Delta/T}}{(2\pi)^{3/2}\hbar^3} \tag{2-8$'$}$$

By differentiating relation 2-8 we find the entropy of the roton gas

$$S_r = -\frac{\partial F_r}{\partial T} = k\mathfrak{N}_r\left(\frac{\Delta}{T} + \frac{3}{2}\right) \tag{2-9}$$

Let us furthermore calculate the specific heat of the roton gas

$$C_r = T\frac{\partial S_r}{\partial T} = k\mathfrak{N}_r\left(\frac{\Delta^2}{T^2} + \frac{\Delta}{T} + \frac{3}{4}\right) \tag{2-10}$$

By summing the results found in Eqs. 2-6, 2-7, 2-9, and 2-10 we find expressions for the entropy and specific heat per unit volume of helium II

$$S = S_r + S_{ph} = k\mathfrak{N}_r\left(\frac{\Delta}{T} + \frac{3}{2}\right) + \frac{16\pi^5 k}{45}\left(\frac{kT}{2\pi\hbar c}\right)^3 \tag{2-11}$$

$$C = C_r + C_{ph} = k\mathfrak{N}_r\left(\frac{\Delta^2}{T^2} + \frac{\Delta}{T} + \frac{3}{4}\right) + \frac{16\pi^5 k}{15}\left(\frac{kT}{2\pi\hbar c}\right)^3 \tag{2-12}$$

Analogous calculations to the ones shown here may be carried out for the case of nonzero relative velocity $\mathbf{w} = \mathbf{v}_n - \mathbf{v}_s$. We shall not give the details of the calculations, which are not complicated, but only write down the final results obtained in this case:

$$\overline{F}_{ph} = F_{ph}(1 - w^2/c^2)^{-2} \tag{2-13}$$

$$\overline{F}_r = F_r \frac{kT}{p_0 w} \sinh \frac{p_0 w}{kT} \tag{2-14}$$

$$\overline{S}_{ph} = S_{ph} (1 - w^2/c^2)^{-2} \tag{2-15}$$

$$\overline{C}_{ph} = C_{ph} (1 - w^2/c^2)^{-2} \tag{2-16}$$

$$\overline{S}_r = S_r \frac{kT}{p_0 w} \sinh \frac{p_0 w}{kT} + F_r \frac{1}{T} \times \left(\frac{kT}{p_0 w} \sinh \frac{p_0 w}{kT} - \cosh \frac{p_0 w}{kT} \right) \tag{2-17}$$

$$\overline{C}_r = C_r \frac{kT}{p_0 w} \sinh \frac{p_0 w}{kT} + F_r \frac{1}{T} \frac{p_0 w}{kT} \sinh \frac{p_0 w}{kT} \tag{2-18}$$

$w = |\mathbf{v}_n - \mathbf{v}_s|$ is the relative velocity of the normal and superfluid motions. The unbarred quantities denote thermodynamic functions of helium II at rest.

3 THE NORMAL DENSITY

The momentum per unit volume of helium II in the frame of reference moving with the superfluid part can, by Eq. 1-9, be written

$$\mathbf{j}_0 = \mathbf{j} - \rho \mathbf{v}_s = \rho_n (\mathbf{v}_n - \mathbf{v}_s) \tag{2-19}$$

On the other hand this momentum can be represented in the form of an integral

$$\int \mathbf{p} n(\epsilon + \mathbf{p} \cdot \mathbf{v}_s - \mathbf{p} \cdot \mathbf{v}_n)\, d\tau_p \tag{2-20}$$

taken over all elementary excitations. By comparing Eqs. 2-19 and 2-20 we obtain the relation defining the normal density

$$\rho_n (\mathbf{v}_n - \mathbf{v}_s) = \int \mathbf{p} n(\epsilon + \mathbf{p} \cdot \mathbf{v}_s - \mathbf{p} \cdot \mathbf{v}_n)\, d\tau_p \tag{2-21}$$

In the case of small values of the difference $\mathbf{v}_n - \mathbf{v}_s$ one can expand the distribution function n in powers of this difference. The zero'th-order term on the right hand side of Eq. 2-21 is zero. The first-order term in the expansion gives

$$\rho_n = -\frac{1}{3kT} \int p^2 n'\, d\tau_p \tag{2-22}$$

Formula 2-21 holds for small values of the difference $\mathbf{v}_n - \mathbf{v}_s$.

Let us first calculate the phonon part of the normal density. Inserting the distribution function, Eq. 1-13, into Eq. 2-21 we obtain

$$\rho_{n\,ph}\,(\mathbf{v}_n - \mathbf{v}_s) = \int \mathbf{p}\left[\exp\left(\frac{\epsilon + \mathbf{p}\cdot\mathbf{v}_s - \mathbf{p}\cdot\mathbf{v}_n}{kT}\right) - 1\right]^{-1} d\tau_{\mathbf{p}} \tag{2-23}$$

By performing the simple integration and dividing both sides by the factor $\mathbf{v}_n - \mathbf{v}_s$ we find

$$\rho_{n\,ph} = \frac{4}{3}\frac{E_{ph}}{c^2}\,(1 - w^2/c^2)^{-3} \tag{2-24}$$

For rotons we may perform an analogous calculation. We get

$$\rho_{nr}\,(\mathbf{v}_n - \mathbf{v}_s) = \int \mathbf{p}\,\exp\left(-\frac{\epsilon + \mathbf{p}\cdot\mathbf{v}_s - \mathbf{p}\cdot\mathbf{v}_n}{kT}\right) d\tau_{\mathbf{p}}$$

from which we easily find

$$\rho_{nr} = \mathfrak{N}_r\,\frac{kT}{w^2}\left(\cosh\frac{p_0 w}{kT} - \frac{kT}{p_0 w}\sinh\frac{p_0 w}{kT}\right) \tag{2-25}$$

The normal density of helium ρ_n is equal to the sum of $\rho_{n\,ph}$ and ρ_{nr}. According to Eqs. 2-21 and 2-25 we have

$$\rho_n = \mathfrak{N}_r\,\frac{kT}{w^2}\left(\cosh\frac{p_0 w}{kT} - \frac{kT}{p_0 w}\sinh\frac{p_0 w}{kT}\right) + \frac{4}{3}\frac{E_{ph}}{c^2}\,(1 - w^2/c^2)^{-3} \tag{2-26}$$

For small values of the velocity of relative motion of the normal and superfluid components we can neglect the dependence of $\rho_{n\,ph}$ and ρ_{nr} on $w = |\mathbf{v}_n - \mathbf{v}_s|$. In that case formulas 2-24 and 2-25 give

$$\rho_{n\,ph} = \frac{4}{3}\frac{E_{ph}}{c^2} \tag{2-27}$$

$$\rho_{nr} = \frac{p_0^2}{3kT}\,\mathfrak{N}_r \tag{2-28}$$

$$\rho_n = \frac{4}{3}\frac{E_{ph}}{c^2} + \frac{p_0^2}{3kT}\mathfrak{N}_r$$

In the temperature range above 0.8 to 1°K, it is the rotons that contribute most to the thermodynamic functions. However the relative contribution of rotons to these functions decreases rapidly with decreasing temperature. This is because the roton contributions decrease exponentially with temperature whereas the phonon contributions go as T^3.

In practice, as we already mentioned, one need not in most problems take into account the dependence of thermodynamic quantities on the relative velocity $\mathbf{w} = \mathbf{v}_n - \mathbf{v}_s$. This is because the ratios w/c and wp_0/kT are very small in the region of velocities where the phenomenon of superfluidity is observable. It is only in treating the problem of the propagation of large amplitude sound waves that one may not neglect this dependence. Finally let us remark that since $c > kT/p_0$, the dependence on w is more pronounced for rotons than for phonons.

3

THE QUANTIZATION OF THE MOTION OF THE LIQUID

A classical liquid may be described by specifying the density ρ and the current vector $\mathbf{j}$, which are defined in the following manner:

$$\rho = \sum m_\alpha \, \delta(\mathbf{R} - \mathbf{r}_\alpha) \tag{3-1}$$

$$\mathbf{j} = \sum \mathbf{p}_\alpha \, \delta(\mathbf{R} - \mathbf{r}_\alpha) = \sum m_\alpha \mathbf{v}_\alpha \, \delta(\mathbf{R} - \mathbf{r}_\alpha) \tag{3-2}$$

The sum is taken over the particles of the system, and $\mathbf{v}_\alpha$ and $\mathbf{p}_\alpha$ are the velocity and momentum of the α-th particle of mass m_α. In order to describe a quantum liquid it is necessary to represent the quantities ρ and $\mathbf{j}$ by operators.

Let us consider one particle of mass m. The operator ρ for this particle is defined in such a manner that its expectation value $\int \psi^*(\mathbf{r}) \rho \, \psi(\mathbf{r}) \, dV$ should be equal to $m\psi^*(\mathbf{R}) \, \psi(\mathbf{R})$. The operator $\rho = m \, \delta(\mathbf{R} - \mathbf{r})$ satisfies this condition. Similarly, for a many-particle system, the operator ρ will have the form 3-1.

In quatum mechanics the current vector is equal to

$$\frac{\hbar}{2i} \{\psi^*(\mathbf{R}) \nabla \psi(\mathbf{R}) - \psi(\mathbf{R}) \nabla \psi^*(\mathbf{R})\} \tag{3-3}$$

Once again, for simplicity let us consider the classical momentum density for one particle $\mathbf{j} = \mathbf{p} \, \delta(\mathbf{R} - \mathbf{r})$. The corresponding symmetrized quantum operator is equal to

$$\mathbf{j} = \tfrac{1}{2}\{\mathbf{p} \, \delta(\mathbf{R} - \mathbf{r}) + \delta(\mathbf{R} - \mathbf{r})\mathbf{p}\} \tag{3-4}$$

where $\mathbf{p} = (\hbar/i)\nabla$ is the momentum operator (∇ acts on the coordinate $\mathbf{r}$). We may convince ourselves that the expectation value of the operator Eq. 3-4 is equal to Eq. 3-3. Indeed we have

$$\int \psi^* \mathbf{j} \psi \, dV = \frac{\hbar}{2i} \int \psi^* \nabla \, \delta(\mathbf{R} - \mathbf{r}) \psi \, dV + \frac{\hbar}{2i} \int \psi^* \, \delta(\mathbf{R} - \mathbf{r}) \nabla \psi \, dV$$

which after an integration by parts in the first term yields

$$\int \psi^* \mathbf{j} \psi \, dV = \frac{\hbar}{2i} \int \{ -\psi \, \delta(\mathbf{R} - \mathbf{r}) \nabla \psi^* + \psi^* \, \delta(\mathbf{R} - \mathbf{r}) \nabla \psi \}$$

$$= \frac{\hbar}{2i} \{ \psi^* \nabla \psi - \psi \nabla \psi^* \}$$

which is just expression 3-3.

In the general case of an arbitrary many-particle system the current operator is equal to the sum of expressions of the form 3-4 over all the particles

$$\mathbf{j} = \tfrac{1}{2} \sum [\mathbf{p}_\alpha \, \delta(\mathbf{R} - \mathbf{r}_\alpha) + \delta(\mathbf{R} - \mathbf{r}_\alpha) \mathbf{p}_\alpha] \qquad \mathbf{p}_\alpha = \frac{\hbar}{i} \nabla_\alpha \tag{3-5}$$

In hydrodynamics, along with the current, one uses the velocity $\mathbf{v}$. The corresponding operator is equal to

$$\mathbf{v} = \frac{1}{2} \left(\frac{1}{\rho} \mathbf{j} + \mathbf{j} \frac{1}{\rho} \right) \tag{3-6}$$

The commutation relations between these operators can be obtained directly by calculating the commutators. As an example, we shall carry out the calculation of the commutator of $\mathbf{j}$ with ρ. According to Eqs. 3-1 and 3-5 we have

$$\mathbf{j}_1 \rho_2 - \rho_2 \mathbf{j}_1 = \sum \frac{m_\alpha \hbar}{2i} \{ [\nabla_\alpha \, \delta(\mathbf{r}_\alpha - \mathbf{R}_1) + \delta(\mathbf{r}_\alpha - \mathbf{R}_1) \nabla_\alpha]$$

$$\times \, \delta(\mathbf{r}_\alpha - \mathbf{R}_2) - \delta(\mathbf{r}_\alpha - \mathbf{R}_2)$$

$$\times \, [\nabla_\alpha \, \delta(\mathbf{r}_\alpha - \mathbf{R}_1) + \delta(\mathbf{r}_\alpha - \mathbf{R}_1) \nabla_\alpha] \}$$

$$= \sum \frac{m_\alpha \hbar}{i} \, \delta(\mathbf{r}_\alpha - \mathbf{R}_1) \nabla \, \delta(\mathbf{R}_1 - \mathbf{R}_2)$$

$$= \frac{\hbar}{i} \rho_1 \nabla \, \delta(\mathbf{R}_1 - \mathbf{R}_2) \tag{3-7}$$

In an analogous manner one may obtain the other commutation relations (we denote the values of the operators at points $\mathbf{R}_1$ and $\mathbf{R}_2$, respectively, by the indices 1 and 2).

$$\rho_1\rho_2 - \rho_2\rho_1 = 0 \tag{3-8}$$

$$\mathbf{v}_1\rho_2 - \rho_2\mathbf{v}_1 = \frac{\hbar}{i}\nabla\,\delta(\mathbf{R}_1 - \mathbf{R}_2) \tag{3-9}$$

$$v_{1i}v_{2k} - v_{2k}v_{1i} = \frac{\hbar}{i}\,\delta(\mathbf{R}_1 - \mathbf{R}_2)\,\frac{1}{\rho_1}\,(\text{curl }\mathbf{v})_{ik} \tag{3-10}$$

One can obtain one more relation by taking the curl of both sides of Eq. 3-9,

$$\text{curl }\mathbf{v}_1\rho_2 - \rho_2\,\text{curl }\mathbf{v}_1 = 0 \tag{3-11}$$

By applying the commutation relations to the macroscopic motion of the liquid, we obtain the equations of hydrodynamics in operator form. The energy density of a microscopic system is equal to

$$\frac{\rho\mathbf{v}^2}{2} + E(\rho) \tag{3-12}$$

The macroscopic character of our description is apparent in the second term of Eq. 3-12 which expresses the internal energy $E(\rho)$ as a function of the density ρ, only. The corresponding quantum operator has the form

$$\frac{\mathbf{v}\cdot\rho\mathbf{v}}{2} + E(\rho) \tag{3-13}$$

and the Hamiltonian function of the system is equal to the integral

$$H = \int\left[\frac{\mathbf{v}\cdot\rho\mathbf{v}}{2} + E(\rho)\right]dV \tag{3-14}$$

The equations of hydrodynamics determine the time derivatives of the variables ρ and $\mathbf{v}$. In order to calculate these quantities we take the commutator of the Hamiltonian with the corresponding functions.

We thus have

$$\dot{\rho} = \frac{i}{\hbar}(H\rho - \rho H)$$

Inserting the Hamiltonian operator in the form Eq. 3-13 into this equation we obtain, after a simple calculation, the continuity equation in operator form

$$\dot{\rho} + \text{div}\, \frac{\rho \mathbf{v} + \mathbf{v}\rho}{2} = 0 \tag{3-15}$$

In an analogous manner we can find the time derivative of the velocity

$$\dot{\mathbf{v}} = \frac{i}{\hbar}\,(H\mathbf{v} - \mathbf{v}H)$$

from which the operator equation of motion follows:

$$\dot{v}_i + \left(v_k \frac{\partial v_i}{\partial x_k} + \frac{\partial v_i}{\partial x_k} v_k \right) = -\frac{1}{\rho}\,\nabla p \tag{3-16}$$

where the pressure p is equal to $\rho^2 (\partial/\partial\rho)(E/\rho)$ as in a classical liquid.

4

THE RELATION BETWEEN THE EXCITATION SPECTRUM AND THE STRUCTURE FACTOR OF LIQUID HELIUM II

There exists an intimate connection between the energy spectrum and the liquid structure factor (Feynman[5]). Let us investigate an excited state of a liquid made up of Bose particles. The wave function of this state will be assumed to be a symmetric sum

$$\sum_a f(\mathbf{r}_a)\cdot\Phi \tag{4-1}$$

taken over all the atoms of the system. $f(\mathbf{r}_a)$ is some function of the radius vector of the a-th atom and Φ, the wave function of the ground state, which depends on the coordinates of all the atoms. In order to find the function $f(\mathbf{r}_a)$ we shall apply a variational method.

Let us write the Hamiltonian of the system as

$$\hat{H} = -\frac{\hbar^2}{2m}\sum_a \nabla_a^2 + V - E_0 \tag{4-2}$$

where V is the potential energy of the system, and energies are measured relative to the ground-state energy E_0. The ground-state wave function then satisfies the equation

$$\hat{H}\Phi = 0 \tag{4-3}$$

We introduce the function F defined by

$$\psi = F\Phi \tag{4-4}$$

By applying the Hamiltonian, Eq. 4-2, to the function 4-4 and by using Eq. 4-3, we obtain

$$\hat{H}\psi = \hat{H}(F\Phi) = -\frac{\hbar^2}{2m}\sum_a (\Phi\nabla_a^2 F + 2\nabla_a\Phi\nabla_a F)$$

$$= -\Phi^{-1}\frac{\hbar^2}{2m}\sum_a \nabla_a(\rho_{\mathfrak{N}}\nabla_a F) \tag{4-5}$$

Here $\rho_{\mathfrak{N}} = \Phi^2$ is the probability density for the ground state.† It determines the probability that the system will be in any configuration $\mathbf{r}^{\mathfrak{N}}$($\mathbf{r}^{\mathfrak{N}}$ denotes the set of radius vectors of all $\mathfrak{N}$ atoms in the system). The energy of the system can be found as the minimum of the quantity

$$\mathcal{E} = \int \psi^*\hat{H}\psi\, d^{\mathfrak{N}}\mathbf{r} = \frac{\hbar^2}{2m}\sum_a \int (\nabla_a F^*\nabla_a F)\,\rho_N\, d^{\mathfrak{N}}\mathbf{r} \tag{4-6}$$

with the subsidiary condition that the normalization integral

$$\mathcal{I} = \int \psi^*\psi\, d^{\mathfrak{N}}\mathbf{r} = \int F^*F\rho_{\mathfrak{N}}\, d^{\mathfrak{N}}\mathbf{r} \tag{4-7}$$

has a fixed value. The energy E will then equal $\mathcal{E}/\mathcal{I}$. According to Eq. 4-2 we express the function F as a sum

$$F = \sum_a f(\mathbf{r}_a) \tag{4-8}$$

over all the atoms of the system. We insert this expression into the normalization integral, Eq. 4-7,

$$\mathcal{I} = \int \sum_a \sum_b f^*(\mathbf{r}_a)f(\mathbf{r}_b)\rho_{\mathfrak{N}}\, d^{\mathfrak{N}}\mathbf{r} \tag{4-9}$$

and fix two values a and b of the particle indices. We then perform

†Since the ground-state wave function has no nodes we may take it to be real.

the integrations over all other particle coordinates in Eq. 4-9; in this manner we obtain

$$\mathcal{I} = \int f^*(\mathbf{r}_1) f(\mathbf{r}_2) \rho_2(\mathbf{r}_1, \mathbf{r}_2)\ d^3\mathbf{r}_1\ d^3\mathbf{r}_2 \tag{4-10}$$

where ρ_2 is the probability of finding one atom at $\mathbf{r}_1$ and another at $\mathbf{r}_2$. In an analogous manner we may integrate $\rho_{\mathfrak{N}}$ over all coordinates except one and obtain the function $\rho_1(\mathbf{r})$ which gives the probability of finding an atom at $\mathbf{r}$ in the ground state of the liquid. This quantity obviously does not depend on $\mathbf{r}$ and is, therefore, equal to some number, ρ_0. As for the function $\rho_2(\mathbf{r}_1, \mathbf{r}_2)$, since the liquid is homogeneous and isotropic, we can write it in the form $\rho_2 = \rho_0\, p(\mathbf{r}_1 - \mathbf{r}_2)$. We can thus rewrite the normalization integral in the form

$$\mathcal{I} = \rho_0 \int f^*(\mathbf{r}_1) f(\mathbf{r}_2)\, p(\mathbf{r}_1 - \mathbf{r}_2)\ d^3\mathbf{r}_1\ d^3\mathbf{r}_2 \tag{4-11}$$

If we now take expression 4-6 for the energy $\mathcal{E}$ and insert the function F written in the form 4-8, we can integrate over all particle coordinates except one and obtain

$$\mathcal{E} = \frac{\hbar^2}{2m} \sum_a \int \nabla_a f^*(\mathbf{r}_a) \nabla_a f(\mathbf{r}_a) \rho_N\ d^{\mathfrak{N}}\mathbf{r}$$

$$= \rho_0 \frac{\hbar^2}{2m} \int \nabla f^*(\mathbf{r}) \nabla f(\mathbf{r})\ d^3\mathbf{r} \tag{4-12}$$

We now choose the function $f(\mathbf{r})$ in such a way that the ratio $\mathcal{E}/\mathcal{I}$ has a minimum value. Taking the variation of $\mathcal{E}/\mathcal{I}$ with respect to f^*, we have the equation

$$\mathcal{E} \int \rho(\mathbf{r}_1 - \mathbf{r}_2) f(\mathbf{r}_2)\ d^3\mathbf{r}_2 = -\frac{\hbar^2}{2m} \nabla^2 f(\mathbf{r}_1) \tag{4-13}$$

where we used Eqs. 4-11 and 4-12. This equation has the solution†

$$f(\mathbf{r}) = \exp(i\mathbf{k}\cdot\mathbf{r}) \tag{4-14}$$

The value of the energy corresponding to this solution is

$$E(k) = \frac{\hbar^2 k^2}{2mS(k)} \tag{4-15}$$

†A function of this form was first proposed by Bijl.[5] For small values of the wave vector **k**, Bijl wrote the wave function of an excited state in the form $\Sigma_a\, e^{i\mathbf{k}\cdot\mathbf{r}_a}\, \Phi$.

where S(**k**) is the Fourier transform of the correlation function

$$S(\mathbf{k}) = \int p(\mathbf{r}) \exp(i\mathbf{k}\mathbf{r})\, d^3\mathbf{r} \tag{4-16}$$

As can be seen from its relation to the correlation function p(**r**), the function S(**k**) is the liquid structure factor, which characterizes the interaction of the liquid at zero temperature with various types of external probes (neutron scattering, scattering of γ-rays, etc.). The structure factor S(**k**) cannot be calculated. It is, however, well known experimentally. The most characteristic properties of this function can be obtained from very simple considerations. For large values of k the function S(**k**) tends to unity and the correlation function becomes a δ-function. The function S(**k**) has a maximum at values of k of the order of $2\pi/a$ (a is the interparticle distance) which corresponds to a maximum of the function p(**r**) for r of the order of a, namely when the two atoms are at the average interparticle distance from one another. Finally, for small values of k the function S(**k**) tends to zero linearly, corresponding to the fact that at large distances correlations disappear.

In Fig. 2 we show the form of the function S(**k**) and the resulting form for E(k). The linear part, for small k, corresponds to the phonon part of the spectrum. Near its minimum, the curve E(k) has the form

$$E(k) = \Delta + \frac{\hbar^2}{2\mu}(k - k_0)^2$$

Finally for $k \to \infty$, $E(k) = k^2/2m$.

We now present a different but instructive derivation of the Feynman formulas, based on the hydrodynamics of the quantum liquid.[6] We write the Hamiltonian of the quantum liquid (see Eq. 3-14) as

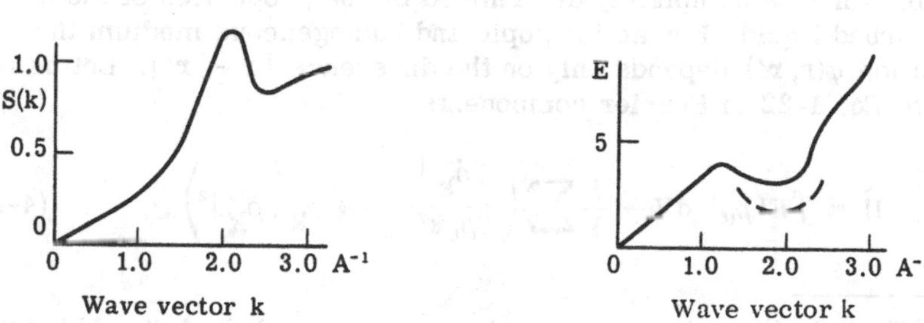

Figure 2. The structure factor in liquid helium S(**k**) and the resulting energy spectrum $E(k) \cdot 2m\hbar^2$.

$$\hat{H} = \int \left[\frac{\mathbf{v} \cdot \rho \mathbf{v}}{2} + E(\rho) \right] dV \tag{4-17}$$

For weakly excited states of the liquid we can write the density in the form

$$\rho(\mathbf{r}) = \rho_0 + \rho_1 \qquad (\rho_1 \ll \rho_0) \tag{4-18}$$

where ρ_0 = const. We further expand ρ_1 and $\mathbf{v}$ in Fourier series

$$\rho_1 = \frac{1}{\sqrt{V}} \sum_{\mathbf{k}} (\rho_{\mathbf{r}} \, e^{i\mathbf{k}\cdot\mathbf{r}} + \text{c.c.}) \tag{4-19}$$

$$\mathbf{v} = \frac{1}{\sqrt{V}} \sum_{\mathbf{k}} (\mathbf{v}_{\mathbf{k}} \, e^{i\mathbf{k}\cdot\mathbf{r}} + \text{c.c.}) \tag{4-20}$$

From the equation of continuity

$$\dot{\rho} + \operatorname{div} \rho \mathbf{v} = 0$$

we obtain the relation between $\mathbf{v}_{\mathbf{k}}$ and $\rho_{\mathbf{k}}$

$$\mathbf{v}_{\mathbf{k}} = i \frac{\dot{\rho}_{\mathbf{k}}}{\rho_0} \frac{\mathbf{k}}{k^2} \tag{4-21}$$

We now expand the Hamiltonian $\hat{H}$ in a series in ρ_1 and $\mathbf{v}$ and limit ourselves to the quadratic terms,† obtaining

$$\hat{H} = \int E(\rho_0) \, dV + \tfrac{1}{2} \int \rho_0 \mathbf{v}^2 \, dV + \tfrac{1}{2} \int \varphi(\mathbf{r}, \mathbf{r}') \rho_1(\mathbf{r}') \rho_1(\mathbf{r}) \, dV \, dV' \tag{4-22}$$

Here $\varphi(\mathbf{r}, \mathbf{r}')$ is the second functional derivative of the energy density $E(\rho)$, which is completely determined by the properties of the unperturbed liquid. For an isotropic and homogeneous medium the function $\varphi(\mathbf{r}, \mathbf{r}')$ depends only on the difference $|\mathbf{r} - \mathbf{r}'|$. Let us rewrite Eq. 4-22 in Fourier components

$$\hat{H} = \int E(\rho_0) \, dV + \frac{1}{V} \sum \left(\frac{|\dot{\rho}_{\mathbf{k}}|^2}{2\rho_0 k^2} + \tfrac{1}{2} \varphi_{\mathbf{k}} |\rho_{\mathbf{k}}|^2 \right) \tag{4-23}$$

---

†The term linear in ρ_1 in expansion 4-22 is zero. Indeed, it would have the form $\int \psi(\mathbf{r}) \rho_1(\mathbf{r}) \, dV$. Because of the isotropy and the homogeneity of the medium, $\psi(\mathbf{r})$ = const. and the integral $\int \rho_1(\mathbf{r}) \, dV$ vanishes.

This expression has the form of a sum of Hamiltonians for harmonic oscillators, with frequencies

$$\omega^2(\mathbf{k}) = k^2 \varphi_{\mathbf{k}} \rho_0 \qquad (4\text{-}24)$$

Thus the weakly excited states of the liquid are represented by a set of elementary excitations, each one of which is described by the equation for a harmonic oscillator. According to the general laws of quantum mechanics the energy of an oscillator is related to its frequency by the equation

$$E_{\mathbf{k}} = \hbar\omega_{\mathbf{k}}(n + \tfrac{1}{2}) \qquad n = 0, 1, 2 \ldots \qquad (4\text{-}25)$$

The ground-state energy of the liquid is thus equal to

$$E_0 = \int E(\rho_0)\, dV + \sum \frac{\hbar\omega_{\mathbf{k}}}{2} \qquad (4\text{-}26)$$

where the second term in Eq. 4-26 is the zero-point energy of the oscillators. By comparing Eqs. 4-26 and 4-23 we find

$$\frac{\hbar\omega_{\mathbf{k}}}{2} = \frac{1}{V}\left(\frac{\overline{|\dot{\rho}_{\mathbf{k}}|^2}}{2\rho_0 k^2} + \tfrac{1}{2}\varphi_{\mathbf{k}}\overline{|\rho_{\mathbf{k}}|^2}\right) = \frac{1}{V}\varphi_{\mathbf{k}}\overline{|\rho_{\mathbf{k}}|^2} \qquad (4\text{-}27)$$

Finally from Eqs. 4-24 and 4-27 we again obtain formula 4-15 for the energy of an elementary excitation

$$\epsilon_{\mathbf{k}} = \hbar\omega_{\mathbf{k}} = \frac{k^2\hbar^2}{2mS(\mathbf{k})} \qquad (4\text{-}28)$$

where $S(\mathbf{k}) = |\rho_{\mathbf{k}}|^2/Vm\rho_0$ is the Fourier transform of the density-correlation function

$$S(\mathbf{r} - \mathbf{r}') = \frac{\overline{(n(\mathbf{r}) - \bar{n})(n(\mathbf{r}') - \bar{n})}}{\bar{n}} \qquad (4\text{-}29)$$

Here $n(\mathbf{r}) = \rho(\mathbf{r})/m$ is the number density of the particles.

The above hydrodynamic derivation is valid only for wave vectors k less than $1/a$ (a is the interatomic distance)—that is, for a situation in which the liquid can be treated as a continuous medium. Therefore formula 4-15 is only valid for small values of the wave vector. Since it gives correct results in the limit $k \to \infty$, namely, in the limit of free particles, one can use it as an interpolation formula in the momentum region $k \sim 1/a$. It is natural to generalize the hydrodynamic derivation to the case $T \neq 0$. In this case the energy $E(\rho,S)$ depends on two variables ρ and S, and in the Fourier expansion there appear two kinds of oscillators. The first kind are the ones previously considered and the second kind are oscillators of "second sound." For small values of the momentum, the excitation energy of the second kind of oscillator is given by the formula

$$E_{\mathbf{k}} = \hbar u_2 k \tag{4-30}$$

where u_2 is the velocity of second sound. We thus obtain excitations that are quanta of second sound.† It must however be noted that one may only speak of second sound quanta if their wavelength is longer than the mean free path of the ordinary excitations—that is, if the two-fluid hydrodynamics holds. Because of this, the contribution of second sound quanta to any physical effect is limited to wavelengths longer than the mean free path of the excitations (phonons and rotons) and is, therefore, negligibly small.

†For more details on second sound, see Chap. 10.

5

THE ENERGY SPECTRUM OF A WEAKLY NONIDEAL BOSE GAS

The energy spectrum of real helium can of course not be determined theoretically. It is therefore interesting to investigate some model whose energy spectrum has properties similar to those described above.

Let us consider a nonideal Bose gas of particles with zero spin. The excitation spectrum can be obtained exactly in the case of weak interactions when perturbation theory is applicable (Bogoliubov[7]).

The Hamiltonian of the system can be written in second quantization

$$\hat{H} = \sum \frac{p^2}{2m} a_p^\dagger a_p + \frac{1}{2V} \sum \mathcal{V}_{p_1 p_2; p_3 p_4} a_{p_1}^\dagger a_{p_2}^\dagger a_{p_3} a_{p_4} \qquad (5\text{-}1)$$

where the operators $a_p^\dagger$ and a_p satisfy the well-known commutation relations

$$a_p a_{p'}^\dagger - a_{p'}^\dagger a_p = \delta_{pp'} \qquad a_p a_{p'} - a_{p'} a_p = 0$$
$$a_p^\dagger a_{p'}^\dagger - a_{p'}^\dagger a_p^\dagger = 0 \qquad (5\text{-}2)$$

Let us consider the system to be near zero temperature. Then, because of Bose condensation, almost all the particles will be in the condensate—that is, will have zero energy ($p = 0$).

We shall be interested in the low-lying excited states (small p). Therefore we can replace the interaction matrix element $\mathcal{V}_{p_1 p_2; p_3 p_4}$

by a constant which we shall denote by α. The next simplification which occurs for weakly excited states is due to the fact that almost all the particles are in the condensate. Therefore the total number of particles N is equal to

$$N = N_0 + \sum_p N_p \tag{5-3}$$

where $\Sigma N_p \ll N_0$. Since the number of condensed particles is large, we can treat these particles classically and replace everywhere the operators $a_0^\dagger$ and a_0 by $(N_0)^{1/2}$.

The largest contribution to the interaction energy comes from the interaction of condensate particles among themselves $O + O \rightarrow O + O$, and the interaction of excited particles (particles not in the condensate) with condensate particles.

We can write these terms in the form

$$\frac{N_0^2}{2V}\alpha + \frac{N_0}{V}\alpha \sum \left(a_p a_{-p} + a_p^\dagger a_{-p}^\dagger + 2a_p^\dagger a_p + 2a_{-p}^\dagger a_{-p}\right) \tag{5-4}$$

We further express N_0 in terms of N and $\Sigma a_p^\dagger a_p$ with the help of Eq. 5-3 and insert into Eq. 5-4, obtaining

$$\frac{N^2}{2V}\alpha + \frac{N}{V}\alpha \sum \left(a_p a_{-p} + a_p^\dagger a_{-p}^\dagger + a_p^\dagger a_p + a_{-p}^\dagger a_{-p}\right) \tag{5-5}$$

To the accuracy which we are interested in, we can write the Hamiltonian in the form

$$\hat{H} = \sum \left(\alpha N + \frac{p^2}{2m}\right)\left(a_p^\dagger a_p + a_{-p}^\dagger a_{-p}\right) + \frac{N\alpha}{V} \sum \left(a_p a_{-p} + a_p^\dagger a_{-p}^\dagger\right) \tag{5-6}$$

This Hamiltonian which is quadratic in the operators a_p and $a_p^\dagger$ can be diagonalized by performing a Bogoliubov transformation. We introduce the new operators $\alpha_p^\dagger$ and α_p related to $a_p^\dagger$ and a_p by the expression

$$\begin{aligned} a_p &= u_p \alpha_p + v_p \alpha_{-p}^\dagger \\ a_p^\dagger &= u_p \alpha_p^\dagger + v_p \alpha_{-p} \end{aligned} \tag{5-7}$$

From the requirement that the new operators satisfy the usual Bose commutation relations it follows that

$$u_p^2 - v_p^2 = 1 \tag{5-8}$$

Inserting expression 5-7 into the Hamiltonian, Eq. 5-6, we obtain

$$\begin{aligned}\hat{H} = \sum_{p>0} &\left\{\left(\frac{p^2}{2m} + \frac{N\alpha}{V}\right)\left(u_p^2 + v_p^2\right) + 2\,\frac{N\alpha}{V}\,u_p v_p\right\} \\ &\times \left(\alpha_p^\dagger \alpha_p + \alpha_{-p}^\dagger \alpha_{-p}\right) \\ &+ \sum_{p>0}\left\{\left(\frac{p^2}{2m} + \frac{N\alpha}{V}\right) 2u_p v_p + \frac{N\alpha}{V}\left(u_p^2 + v_p^2\right)\right\} \\ &+ \left(\alpha_p^\dagger \alpha_{-p}^\dagger + \alpha_p \alpha_{-p}\right) \\ &+ \sum_{p>0}\left\{2\left(\frac{p^2}{2m} + \frac{N\alpha}{V}\right) v_p^2 + 2\,\frac{N\alpha}{V}\,u_p v_p\right\} + \frac{\alpha N^2}{2V}\end{aligned} \tag{5-9}$$

In order that the new Hamiltonian be diagonal it is necessary that terms of the form $\alpha_p^\dagger \alpha_{-p}^\dagger$ be absent; consequently

$$\left(\frac{p^2}{2m} + \frac{N\alpha}{V}\right) 2u_p v_p + \frac{N\alpha}{V}\left(u_p^2 + v_p^2\right) = 0 \tag{5-10}$$

From Eqs. 5-8 and 5-10 we can find the coefficients u_p and v_p and insert them into Eqs. 5-9. We finally obtain

$$\hat{H} = E_0 + \sum_{p>0} \sqrt{\left(\frac{p^2}{2m} + \frac{N\alpha}{V}\right)^2 - \left(\frac{N\alpha}{V}\right)^2}\left(\alpha_p^\dagger \alpha_p + \alpha_{-p}^\dagger \alpha_{-p}\right) \tag{5-11}$$

where the "zero-point" energy E_0 is equal to

$$E_0 = \frac{\alpha N^2}{2V} - \sum_{p>0}\left\{\left(\frac{p^2}{2m} + \frac{N\alpha}{V}\right) - \sqrt{\left(\frac{p^2}{2m} + \frac{N\alpha}{V}\right)^2 - \left(\frac{N\alpha}{V}\right)^2}\right\} \tag{5-12}$$

The result we have obtained is quite remarkable. From Eq. 5-11 we see that the weakly excited states of the system are superpositions of elementary excitations with energy spectrum

$$\epsilon_p = \sqrt{\left(\frac{p^2}{2m} + \frac{N\alpha}{V}\right)^2 - \left(\frac{N\alpha}{V}\right)^2} \tag{5-13}$$

For sufficiently small p

$$p^2 \ll 4m \frac{N\alpha}{V} \tag{5-14}$$

we have

$$\epsilon_{\mathbf{p}} = p \sqrt{\frac{\alpha N}{mV}} \tag{5-15}$$

that is, the elementary excitations are phonons with a velocity equal to the velocity of sound

$$c = \sqrt{\frac{\alpha N}{mV}} \tag{5-16}$$

For larger values of the momentum p, the dispersion law is more complicated; it approaches asymptotically the spectrum of free particles $\epsilon_{\mathbf{p}} = \mathbf{p}^2/2m$. However one must remember that the foregoing analysis was only valid for sufficiently low values of p.

Let us return to expression 5-12 for the ground-state energy E_0. Although the sum in Eq. 5-12 diverges, one can nevertheless obtain finite results, as was shown by Lee and Yang,[8] by expressing the matrix element α in perturbation theory to second order. This gives

$$\alpha^{(2)} = \alpha - \frac{\alpha^2}{V} \sum \frac{m}{p^2} \tag{5-17}$$

If we write α in terms of $\alpha^{(2)}$ and insert into Eq. 5-12 we obtain

$$E_0 = \frac{N^2 \alpha^{(2)}}{2V} - \sum \left\{ \left(\frac{\mathbf{p}^2}{2m} + \frac{N\alpha}{V} \right) - \sqrt{\left(\frac{\mathbf{p}^2}{2m} + \frac{N\alpha}{V} \right)^2 - \left(\frac{N\alpha}{V} \right)^2} - \left(\frac{N\alpha}{V} \right)^2 \frac{m}{\mathbf{p}^2} \right\} \tag{5-18}$$

Let us replace the sum in 5-18 by an integral and perform the simple integration. Finally let us express the matrix element in terms of the more convenient scattering amplitude

$$a = \frac{m\alpha}{4\pi\hbar^2} \tag{5-19}$$

This brings us to the final expression

$$E_0 = \frac{2\pi\hbar^2 aN^2}{mV}\left(1 + \frac{128}{15\sqrt{\pi}}\sqrt{\frac{a^3 N}{V}}\right) \tag{5-20}$$

This represents the first two terms of the expansion of the ground-state energy in powers of $a(N/V)^{1/3}$. We thus see that the foregoing analysis is valid in the case of short-range forces, when the scattering amplitude is small with respect to the interparticles distance.

6

ON THE FORM OF THE ELEMENTARY EXCITATION SPECTRUM IN THE VICINITY OF SINGULAR POINTS[9]

Let us investigate the properties of the spectrum near points where one excitation can decay into two. From the laws of conservation of momentum and energy we then have

$$\epsilon(\mathbf{p}) = \epsilon(\mathbf{q}) + \epsilon(|\mathbf{p} - \mathbf{q}|) \tag{6-1}$$

Here $\mathbf{p}$ is the momentum of the decaying particle, $\mathbf{q}$ and $\mathbf{p} - \mathbf{q}$ the momenta of the created particles. Let us suppose that the spectrum has stable regions. Mathematically this means that up to some momentum $\mathbf{p}_0$ Eq. 6-1 has no solutions, but starting at $\mathbf{p}_0$ it is possible to satisfy Eq. 6-1. In that case it is necessary that the right-hand side of Eq. 6-1 have a minimum as a function of $\mathbf{q}$. The vector $\mathbf{q}$ can be determined by two quantities: its absolute value q and the angle it makes with the vector $\mathbf{p}$. The extremum conditions with respect to the variables q and cos θ are, respectively,

$$\epsilon'(\mathbf{q}) + \epsilon'(|\mathbf{p} - \mathbf{q}|)\,\frac{q - p\cos\theta}{|\mathbf{p} - \mathbf{q}|} = 0 \tag{6-2}$$

$$\epsilon'(|\mathbf{p} - \mathbf{q}|)\,pq = 0 \tag{6-3}$$

The primes denote differentiation of a function with respect to its argument. From condition 6-3 it follows that either $q \to 0$ or $\epsilon'(|\mathbf{p} - \mathbf{q}|) = 0$. Therefore, from Eqs. 6-2 and 6-3 there are two possibilities

$$\text{(a)} \quad q = 0 \qquad \epsilon'(0) = \epsilon'(p)\cos\theta \tag{6-4}$$

$$\text{(b)} \quad \epsilon'(|\mathbf{p} - \mathbf{q}|) = 0 \qquad \epsilon'(q) = 0 \tag{6-5}$$

In case (a) the momentum $\mathbf{q}$ of the emitted particle goes to zero—that is, it is a phonon. From condition 6-4 we see that the emission of a phonon will first occur when $\epsilon'(p) = \pm\epsilon'(0)$. Since in a Bose liquid the long-wavelength excitations are phonons it follows that $\epsilon'(0) = c$, the velocity of sound. Thus at the point in the spectrum where instability first sets in, a phonon is emitted, either in the forward direction ($\theta = 0$) or directed oppositely to the decaying particle ($\theta = \pi$). We have here assumed that the curve $\epsilon(p)$ has a continuous derivative. In the unlikely but in principle possible case where the curve $\epsilon(p)$ has a discontinuity, the phonon will be emitted at an angle θ determined by relation 6-4

$$\cos v = \frac{\epsilon'(0)}{\epsilon'(p)}$$

In case (b) the decay occurs for particles with momenta $\mathbf{q}$ and $\mathbf{p} - \mathbf{q}$, for which the energy ϵ has a minimum. The magnitude of the momentum $\mathbf{q}$ and the angle of emission are determined by the two equations 6-5.

Case c): The extremum of the right-hand side of 6-1 with respect to the angle θ can also be achieved at the boundaries of the interval—that is, at the angles $\theta = 0$ and $\theta = \pi$ [case (c)]. In that case from condition 6-2 it follows that

$$\epsilon'(q) = \pm\epsilon'(p \mp q) \tag{6-6}$$

Thus for such a decay, the excitations are emitted with identical speeds, at angles $\theta = 0$ or $\theta = \pi$ with respect to one another. It can be shown that the three possibilities mentioned above exhaust all the points of instability of the spectrum. We should note that for the excitation spectrum of the weakly interacting Bose gas, Eq. 5-13, the instability appears at the very beginning of the spectrum. Indeed case (a) holds for $q \to 0$, and if $\partial^2\epsilon/\partial q^2 > 0$ then the right-hand side of Eq. 6-1 has a minimum. For the spectrum of liquid helium, the opposite situation holds, because the sign of $\partial^2\epsilon/\partial q^2$ is different in the phonon part of the spectrum; this means that a phonon will not decay into two.

Let us now investigate in detail the spectrum at points where excitations can decay (threshold points). The analysis can be performed by the methods of quantum field theory. It is necessary to study the singularity of the Green's function of an excitation, G(p) near the decay threshold (p is the four-dimensional momentum with components ϵ,p). We assume that the interaction between excitations has a three-particle form. The corresponding vertex part is $\Gamma(p; q; p - q)$. The Green's function G(p) can be expressed in terms of the "zeroth-order" Green's function $G_0(p)$ for "free" excitations, and the vertex part Γ, by means of the Dyson equation

$$G^{-1}(p) = G_0^{-1}(p) + i \int \Gamma(p;\ q;\ p - q)\, G(q)\, G(p - q) \times \Gamma_0(p;\ q;\ p - q) \cdot \frac{d^4 q}{(2\pi)^4} \tag{6-7}$$

where Γ_0 is the vertex part to first order in perturbation theory—that is, some bare interaction.

Case a) The Properties of the Spectrum near the Threshold for Creation of Phonons

Let us suppose that the emitted phonon is stable and that for small q the spectrum has the form ($\alpha > 0$)

$$\omega(q) = cq - \alpha q^3 \tag{6-8}$$

As we shall see below the term containing α in $\omega(q)$ will not be important in what follows. We further suppose that near the threshold ($p = p_c$) the spectrum has the form

$$\epsilon(p) = \epsilon_c + c(p - p_c) + \beta(p - p_c)^2 \tag{6-9}$$

and the singularity appears in the terms of higher order in $(p - p_c)$. Of course our calculations must confirm this last assumption. Now in order that the right-hand side of Eq. 6-1 have a minimum with respect to q it is necessary that β be positive. Indeed for $\cos\theta = 1$ and $p = p_c$ the right-hand side of Eq. 6-1 is equal to

$$\epsilon_c + \beta q^2$$

and has a minimum with respect to q for $\beta > 0$.

For small ω and q the Green's function is equal to

$$G(q) = \frac{a}{\omega^2 - \omega^2(q) + i\delta} \tag{6-10}$$

and is proportional to the phonon propagator.

For $p \approx p_c$ and $\epsilon \approx \epsilon_c$ the Green's function of an excitation has a singularity. We assume that near its zero the function $G^{-1}(p)$ has the form

$$G^{-1}(p) = b^{-1}[c\Delta p + \beta(\Delta p)^2 - \Delta\epsilon - i\delta]$$

$$(\Delta p = p - p_c \quad \Delta\epsilon = \epsilon - \epsilon_c) \tag{6-11}$$

plus terms of higher order which must be determined. The vertex part for a phonon at small q is proportional to q so that we have

$$\Gamma_0(p;\ q;\ p - q) = g_0 q \qquad \Gamma(p;q;\ p - q) = gq \tag{6-12}$$

We are interested in that part of the integral in the Dyson equation, 6-7, which leads to the singularity. A simple analysis of expression 6-7 reveals that the singularity occurs upon integration over small values of ω and q, for which we know all the functions in the integrand. In this way the assumed singular part of integral 6-7 is equal to

$$\text{const} \int \frac{q^4\, dq\, d\cos\theta\, d\omega}{[(cq)^2 - \omega^2 - i\delta][c(\Delta p - q\cos\theta) - \Delta\epsilon + \omega + \beta(\Delta p - q)^2 - i\delta]} \tag{6-13}$$

In this integral the largest contribution comes from angles θ close to zero. We may, therefore, everywhere set $\cos\theta = 1$. We may extend the ω-integral from $-\infty$ to $+\infty$, after which the integration can be performed by taking the residue at $\omega = cq$. As a result we obtain

$$\text{const} \int q^2\, dq \ln(x - 2\beta q\, \Delta p + \beta q^2) \tag{6-14}$$

where $x = c\Delta p - \Delta\epsilon + \beta(\Delta p)^2$.

The last integration over q can be carried out without difficulty by expanding the argument of the logarithm in factors, and we obtain

$$a_1 q_1^3 \ln q_1 + a_2 q_2^3 \ln q_2$$

$$q_{1,2} = \beta\,\Delta p \pm \sqrt{(\beta\,\Delta p)^2 - (\beta - \alpha)x} \tag{6-15}$$

In the immediate vicinity of the pole of G(p) [the zero of $G^{-1}(p)$]—that is, for $x \ll \beta(\Delta p)^2$ we thus obtain the singular part of $G^{-1}(p)$

$$(\Delta p)^3 \ln(-\Delta p) \tag{6-16}$$

Taking Eq. 6-11 into account we have

$$G^{-1}(p) = b^{-1}[c\,\Delta p + \beta(\Delta p)^2 + \gamma(\Delta p)^3 \ln(-\Delta p) - \Delta\epsilon] \tag{6-17}$$

From formula 6-17 it follows that the excitation energy in the vicinity of the phonon emission threshold is equal to

$$\epsilon = \epsilon_c + c(p - p_c) + \beta(p - p_c)^2 + \gamma(p - p_c)^3 \ln(p_c - p) \tag{6-18}$$

For $p > p_c$ there appears a negative imaginary part equal to $-\gamma\pi(\Delta p)^3$ in the energy ϵ, which means that damping occurs. The lifetime is inversely proportional to $(\Delta p)^3$. We may remark that we could have obtained the above result by perturbation theory. This is because the interactions of long-wavelength phonons are always weak. We investigated the case in which the phonon is emitted at an angle $\theta = 0$ with respect to the decaying excitation. For the other possible case, when $\theta = \pi$ only the kinematics of the decay is changed, but the character of the singularity remains the same.

We further investigate case (c) (which is the simpler) and then case (b).

Case c) The Properties of the Spectrum near the Threshold for Decay into Two Excitations with Nonzero but Parallel Momenta

In this case it is physically obvious that the major contribution to Eq. 6-7 comes from momenta q close to the momentum q_0 of the new excitation. The Green's function does not have any singularity near the threshold value q_0 and can be written in the usual form near the pole

$$G(q) = A(\epsilon(q) - \omega - i\delta)^{-1} \tag{6-19}$$

This considerably simplifies the whole ensuing analysis. Let us separate the region of integration in integral 6-7 into two parts, by isolating a small neighborhood close to q_0 and ϵ_0. The integration over this small interval leads to the irregular part of the Green's function near $p = p_c$, as was to be expected. In this small interval we may take the vertex parts Γ and Γ_0 as constants. Thus the irregular part is determined by the integral

$$\int \frac{d^3q\, d\omega}{[\epsilon(q) - \omega - i\delta][\epsilon(|\mathbf{p} - \mathbf{q}|) - \epsilon + \omega - i\delta]} \sim \int \frac{d^3q}{\epsilon(q) + \epsilon(|\mathbf{p} - \mathbf{q}|) - \epsilon} \tag{6-20}$$

The sum $\epsilon(q) + \epsilon(|p - q|)$ has a minimum as a function of q for $p = p_c$, and it can be written in the form

$$\epsilon(q) + \epsilon(|p - q|) \approx \epsilon_c + v_c\, \Delta p + \alpha(q_0 - q)^2 + \beta[(q - q)\cdot p_c]^2/p_c^2 \tag{6-21}$$

Here v_c is the velocity of the created excitation

$$\alpha = \frac{v_c p_c}{2q_0(p_c - q_0)}$$

$$\beta = \frac{1}{2}\left\{\left(\frac{\partial^2 \epsilon}{\partial q^2}\right)_{q=q_0} + \left(\frac{\partial^2 \epsilon}{\partial q^2}\right)_{a=p_c-q_0} - \frac{v_c p_c}{q_0(p_c - q_0)}\right\}$$

Let us introduce new variables $u = |q - q_0|$ and $up = up_c \cos\psi$, after which the integration in Eq. 6-20 can be performed easily, and we get

$$G^{-1} \sim \int \frac{u^2\, du\, d\cos\psi}{v_c\, \Delta p - \Delta\epsilon + \alpha u^2 + \beta u^2 \cos^2\psi} \sim \sqrt{v_c\, \Delta p - \Delta\epsilon} \tag{6-22}$$

Since the point $p = p_c$ and $\epsilon = \epsilon_c$ belongs to the spectrum, we see that for $\Delta p = 0$ and $\Delta\epsilon = 0$ the function $G^{-1}(p)$ should go to zero, and consequently the regular part of $G^{-1}(p)$ should for small values of Δp and $\Delta\epsilon$, have the form $a'\,\Delta p + b'\,\Delta\epsilon$. We thus finally have

$$G^{-1}(p) = A_1^{-1}\left[a\,\Delta p + \Delta\epsilon + b\sqrt{v_c\, \Delta p - \Delta\epsilon}\right] \tag{6-23}$$

The energy of excitation is determined by the equation $G^{-1}(p) = 0$. The solution of this equation for $p < p_c$ has the form

$$\epsilon = \epsilon_c + v_c(p - p_c) - \left(\frac{a + v_c}{b}\right)^2 (p - p_c)^2 \tag{6-24}$$

We must also have the inequality

$$(a + v_c)/b > 0$$

For $p > p_c$ the equation $G^{-1}(p) = 0$ has no real or complex solutions. Thus the curve $\epsilon(p)$ has slope v_c near the threshold and it does not continued beyond the threshold.

Case b) The Decay into Two Excitations Emitted at an Angle with Respect to One Another

In this case, according to Eq. 6-5 the created excitations have an energy corresponding to the minimum of the curve $\epsilon(p)$. Since we are dealing with liquid helium, for which the minimum corresponds to rotons, we shall speak in this case of the decay of the excitation into two rotons with energies

$$\begin{aligned} \epsilon(q) &= \Delta + \frac{(q - p_0)^2}{2\mu} \\ \epsilon(|p - q|) &= \Delta + \frac{(|p - q| - p_0)^2}{2\mu} \end{aligned} \tag{6-25}$$

For the same reasons as previously, the irregular part of the Green's function comes from integration of the Dyson equation over a region of q near p_0. However here the vertex part also turns out to be irregular near the threshold. In order to see what kind of singularity we have here, let us investigate integral 6-20, which is obtained if we first assume in the equation for G^{-1} that Γ is regular (perturbation theory). We can easily perform this integration if we insert Eq. 6-25 and introduce cylindrical coordinates

$$q_z = p_0 \cos\frac{\theta_0}{2} + q'_z \qquad q_x = \left(p_0 \sin\frac{\theta_0}{2} + q'_\rho\right)\cos\varphi$$

$$q_y = \left(p_0 \sin\frac{\theta_0}{2} + q'_\rho\right)\sin\varphi$$

$$\int \frac{dq'_\rho\, dq'_z}{2\Delta - \omega + \frac{1}{\mu}\left(\sin^2\frac{\theta_0}{2}\, q'^2_\rho + \cos^2\frac{\theta_0}{2}\, q'^2_z\right)} \sim \ln(2\Delta - \omega) \tag{6-26}$$

In order to calculate $\Gamma(p;\ p - p_0;\ p_0)$ near $q = p_0$ we would have to solve the corresponding integral equation relating Γ with some bare vertex Γ_0. We shall not discuss this in detail here. We merely remark that by analyzing this equation we are led to summing a geometric series whose terms are powers of integrals of the form Eq. 6-26. In this manner we may show that

$$\Gamma(p;\ p - p_0;\ p_0) \sim \frac{P}{1 + Q \ln \dfrac{2\Delta - \epsilon}{2A}} \tag{6-27}$$

Inserting this Γ into the Dyson equation, 6-7, and carrying out the integration, which is similar to Eq. 6-26 we obtain the irregular part of $G^{-1}(p)$ near the threshold in the form

$$\frac{a}{\ln \dfrac{2\Delta - \epsilon}{\alpha}} \tag{6-28}$$

Since $G^{-1}(p_c) = 0$ we finally have

$$G^{-1}(p) = A^{-1}\left[p - p_c - \frac{a}{\ln \dfrac{2\Delta - \epsilon}{\alpha}}\right] \tag{6-29}$$

From the equation $G^{-1}(p) = 0$ we obtain the spectrum near threshold, for $p < p_c$, with the aid of Eq. 6-29:

$$\epsilon(p) = 2\Delta - \alpha \exp[-a/(p_c - p)] \tag{6-30}$$

Thus the curve $\epsilon(p)$ turns out to have a horizontal tangent at $p = p_c$, with a tangency of infinite order, in this case.

From the presently available neutron scattering data[4] it appears that in superfluid helium this last case obtains—that is, the curve $\epsilon(p)$ ends at a point where decay into two rotons occurs.

7

INTERACTIONS BETWEEN ELEMENTARY EXCITATIONS[10,11]

The beginning of the spectrum is, to a first approximation, linear and represents phonons. However for a certain number of phenomena the dispersion of the phonon part of the spectrum—that is, its departure from a linear dependence, turns out to be of importance. Since this effect is extremely small, it is impossible to determine its magnitude from existing neutron scattering data. It is also not clear whether it can be calculated theoretically by considering a system of interacting phonons. However since the phonon spectrum is stable and the liquid isotropic, one may immediately write down for the initial part of the phonon spectrum the expression

$$\epsilon = cp(1 - \gamma p^2) \tag{7-1}$$

where γ is a positive quantity. One may estimate it by interpolating over the whole energy curve; such an interpolation yields the value

$$\gamma \approx 2.5 \times 10^{37}\ \text{gm}^{-2}\ \text{cm}^{-2}\ \text{sec}^2 \tag{7-2}$$

One must, however, use this estimate with care since it is extremely crude.

We now consider the question of the quantization of the phonon field. Let us represent the density of the liquid $\rho(\mathbf{r})$ and its velocity $\mathbf{v}(\mathbf{r})$ in a series of plane waves

$$\rho(\mathbf{r}) = \rho_0 + \frac{1}{\sqrt{V}} \sum \left(\rho_{\mathbf{k}}\, e^{i\mathbf{k}\cdot\mathbf{r}} + \rho_{\mathbf{k}}^*\, e^{-i\mathbf{k}\cdot\mathbf{r}}\right) \tag{7-3}$$

$$\mathbf{v}(\mathbf{r}) = \frac{1}{\sqrt{V}} \sum \left(\mathbf{v_k} \, e^{i\mathbf{k}\cdot\mathbf{z}} + \mathbf{v_k^*} \, e^{-i\mathbf{k}\cdot\mathbf{r}} \right) \tag{7-4}$$

Here ρ_0 is the equilibrium density, $\mathbf{k} = \mathbf{p}/\hbar$ is the phonon wave vector, which is related to the frequency by the equation

$$\omega = ck \tag{7-5}$$

We further use the commutation relation 3-9

$$\rho(\mathbf{r}_1)\mathbf{v}(\mathbf{r}_2) - \mathbf{v}(\mathbf{r}_2)\rho(\mathbf{r}_1) = \frac{\hbar}{i} \nabla\delta\,(\mathbf{r}_1 - \mathbf{r}_2)$$

from which we get the following relation for the Fourier components, valid when curl $\mathbf{v} = 0$,

$$\rho_{\mathbf{p}}\rho_{\mathbf{q}}^* - \rho_{\mathbf{q}}^*\rho_{\mathbf{p}} = \frac{\rho_0 p}{2c}\, \delta_{\mathbf{pq}} \tag{7-6}$$

$$v_{\mathbf{p}} = \frac{c\mathbf{p}}{\rho_0 p}\, \rho_{\mathbf{p}} \tag{7-7}$$

The Hamiltonian of the system under consideration, namely, liquid helium in a volume V, is equal to

$$\hat{H} = \int_V \{\tfrac{1}{2}\mathbf{v}\cdot\rho\mathbf{v} + E(\rho)\}\, dV \tag{7-8}$$

If we neglect anharmonic terms, we can re-express the equation in terms of the Fourier components of the density

$$\hat{H}_0 = \frac{c^2}{\rho_0} \sum (\rho_{\mathbf{p}}\rho_{\mathbf{p}}^* + \rho_{\mathbf{p}}^*\rho_{\mathbf{p}}) \qquad \langle \hat{H}_0 \rangle = \sum (n_{\mathbf{p}} + \tfrac{1}{2})\hbar\omega \tag{7-9}$$

Here $n_{\mathbf{p}}$ is the number of phonons of momentum $\mathbf{p}$. From Eqs. 7-6 and 7-8 we find the nonzero matrix elements of the Fourier component of the density

$$(\rho_{\mathbf{p}})_{n_{\mathbf{p}},\, n_{\mathbf{p}}+1} = \sqrt{(p\rho_0/2c)(n_{\mathbf{p}} + 1)}\; e^{-i\omega t} \tag{7-10}$$

$$(\rho_{\mathbf{p}}^*)_{n_{\mathbf{p}},\, n_{\mathbf{p}}-1} = \sqrt{(p\rho_0/2c)\, n_{\mathbf{p}}}\; e^{i\omega t} \tag{7-11}$$

Having made these preparatory remarks, let us now turn to the consideration of the various types of interactions between elementary excitations.

PHONON-PHONON SCATTERING

The phonon spectrum, Eq. 7-1, is stable; the decay of a phonon into two or more phonons is not possible, since it would violate the laws of conservation of energy and momentum. The interaction between phonons is due to anharmonic terms in the Hamiltonian, Eq. 7-8. The matrix elements of the density, Eqs. 7-10 and 7-11 are proportional to $\sqrt{p}$ — that is, they depend strongly on the energy of the phonon. At low temperatures, $T < T_\lambda$, the energy of the phonons is much less than the Debye energy. Therefore the phonons interact weakly and we can calculate the possible processes by perturbation theory. The first one of these, phonon-phonon scattering, involves four phonons. If we limit ourselves to terms of fourth order in ρ', we can write the Hamiltonian, Eq. 7-9, in the form

$$H = H_0 + V_3 + V_4 \tag{7-12}$$

where H_0 is the Hamiltonian of the free phonon field, containing terms of second order in ρ', V_3 contains anharmonic terms of third order

$$V_3 = \frac{v\rho' v}{2} + \frac{1}{3!}\frac{\partial}{\partial\rho}\left(\frac{c^2}{\rho}\right)\rho'^3 \tag{7-13}$$

and finally V_4 terms of fourth order in ρ'

$$V_4 = \frac{1}{4!}\frac{\partial^2}{\partial\rho^2}\left(\frac{c^2}{\rho}\right)\rho'^4 \tag{7-14}$$

ρ' is the density variation with respect to its value in the liquid at rest. The transition amplitude for two phonons into two others is obtained in second-order perturbation theory from the terms cubic in ρ' in the energy V_3, and in first-order from the terms quartic in ρ' in the energy V_4,

$$\langle p, p_1 | H | p', p_1' \rangle = \sum \frac{\langle p, p_1 | V_3 | q \rangle \langle q | V_3 | p', p_1' \rangle}{\epsilon(p) + \epsilon(p_1) - \epsilon(q} + \langle p, p_1 | V_4 | p', p_1' \rangle \tag{7-15}$$

The general calculation of the transition amplitudes leads to complicated formulas. However in what follows it turns out that it is sufficient to know the transition amplitude in the limiting case when the momentum of one of the colliding phonos is much less than the other ($p \ll p_1$). We shall limit ourselves to this case. The largest contribution to the sum Eq. 7-15 comes from the intermediate state in which $\mathbf{q} = \mathbf{p} + \mathbf{p}_1$. But if we neglect dispersion in the phonon spectrum we see that the denominator in the sum goes to zero when the angle θ between the momenta $\mathbf{p}$ and $\mathbf{p}_1$ vanishes. We may not, therefore, neglect dispersion. In the special case mentioned above the differential cross section for the process under consideration is

$$d\sigma = \frac{(u+2)^4 p_1^3 p_1' p'}{(16\pi\hbar^2\rho_0)^2\, cp} \frac{\delta\{\epsilon(\mathbf{p}) + \epsilon(\mathbf{p}_1) - \epsilon(\mathbf{p}') - \epsilon(\mathbf{p}_1')\}}{(1-\cos\theta + 3\gamma p_1^2)^2} p'^2\, dp'\, do' \tag{7-16}$$

$$\left(u = \frac{\rho}{c^2}\frac{\partial c^2}{\partial \rho}\right)$$

Since expression Eq. 7-16 has a narrow peak for θ in the forward direction, we may easily carry out the integration needed to obtain the total cross section. This cross section for scattering of a phonon of momentum p with a phonon of momentum p_1, averaged over all angles θ, can finally be written

$$\sigma(\mathbf{p},\mathbf{p}_1) = \frac{\pi(u+2)^4 p_1^4}{(96\pi\hbar^2\rho_0 c)^2\gamma} \qquad (p \ll p_1) \tag{7-17}$$

The dimensionless parameter u is approximately equal to 6. By Eq. 7-16 the effective cross section has a minimum for forward scattering. From the conservation laws it follows that in this case there is no change in direction of any of the phonons but only a rapid exchange of energy between the phonons, which leads to the establishment of energy equilibrium in the phonon gas. The process of establishing this equilibrium plays a fundamental role in all kinetic phenomena in superfluid helium. The exact calculation of the relaxation time characterizing the establishment of energy equilibrium in the phonon gas is not possible, since it is even difficult to formulate the problem exactly. However with the aid of the result given in Eq. 7-17 one may solve this problem in to limiting cases.

In the first case we assume that the phonon distribution function differs from its equilibrium value by some quantity δn in the region of small energies. The collision integral, which characterizes the rate of approach of the distribution function to its equilibrium value, is equal to

$$\mathcal{J}(n) = -\int c\sigma(\mathbf{p},\mathbf{p}_1)\{nn_1(n'+1)(n_1'+1) - n'n_1'(n+1)(n_1+1)\} \times d\mathbf{p}_1(2\pi\hbar)^{-3} \tag{7-18}$$

By assuming the functions n_1, n', and n_1' to be equilibrium functions, we obtain

$$\mathcal{J}(n) = -\delta n \int c\sigma(p,p_1) n^{-1} n' n_1' (n_1 + 1)\, dp_1\, (2\pi\hbar)^{-3} \qquad (7\text{-}19)$$

Thus the relaxation time of a fixed group of phonons is

$$1/t = \int c\sigma(p,p_1)\, n^{-1} n' n_1' (n_1 + 1)\, dp_1\, (2\pi\hbar)^{-3} \qquad (7\text{-}20)$$

For $p \ll p_1$ we may convince ourselves that in the integral, Eq. 7-20, the major contribution comes from phonons with energies of order 6kT. We may, therefore, for simplification, take the equilibrium functions to be Wien functions. Inserting Eq. 7-17 into Eq. 7-21 we finally obtain the relaxation time for low-energy phonons

$$\frac{1}{t_\ell} = \frac{\pi(u + 2)^4 p}{(96\pi\hbar^2\rho_0)\, kT\gamma\, (2\pi\hbar)^3} \int n_1 (n_1 + 1) p_1^4\, dp_1 \qquad (7\text{-}21)$$

which after an elementary integration yields

$$\frac{1}{t_\ell} = \frac{(u + 2)^4\, 6!}{(48\hbar^2\rho_0)^2\, c\gamma\, (2\pi\hbar)^3} \left(\frac{kT}{c}\right)^6 p \qquad (7\text{-}22)$$

We may in a similar manner calculate the relaxation time for high-energy phonons. Assuming $p \gg p_1$, Eq. 7-20 in this case leads to

$$\frac{1}{t_h} = \frac{\pi(u + 2)^4 p^4}{(96\hbar^2\rho_0)^2\, c\gamma\, (2\pi\hbar)^3} \int n_1 (n_1 + 1)\, dp_1 \qquad (7\text{-}23)$$

which again, after an elementary integration, yields

$$\frac{1}{t_h} = \frac{(u + 2)^4 (kT/c)^3 (\pi^2/3)\, p^4}{(48\hbar^2\rho_0)^2\, c\gamma (2\pi\hbar)^3} \qquad (7\text{-}24)$$

The fact that the momentum p_1 is small is automatically taken into account in the integral of Eq. 7-23, since only energies of order 2kT play an important role there. If we insert the numerical values of all the parameters and interpolate between the two limiting cases, we obtain the following formula which is approximately valid at all energies $(x = \epsilon/kT)$:

$$\frac{1}{t_{ph}} = 410^5\, T^7 x(x + 6)^3\ \sec^{-1}$$

A comparison of this time with the time characteristic of other processes shows that energy equilibrium is established very rapidly in the phonon gas. We may, therefore, always assume that the gas is in energy equilibrium.

PHONON-ROTON SCATTERING

Let us calculate the effective cross section for scattering of phonons by rotons. A roton in the presence of the phonon field can be looked upon as a particle in a moving medium. In consequence there appears in the expression for the roton energy an additional term $-\mathbf{p}\cdot\mathbf{v}$ or in symmetric form

$$-\tfrac{1}{2}(\mathbf{p}\cdot\mathbf{v} + \mathbf{v}\cdot\mathbf{p}) \tag{7-26}$$

$\mathbf{p}$ is the momentum operator of the roton and $\mathbf{v}$ the velocity operator of the phonon, Eq. 7-4. The phonon field further changes the density of the medium. If we expand the energy of the roton to second order in the density ρ' we find

$$H_r = H_{ro} + \frac{\partial\Delta}{\partial\rho}\rho' + \frac{1}{2}\left[\frac{\partial^2\Delta}{\partial\rho^2} + \frac{1}{\mu}\left(\frac{\partial p_0}{\partial\rho}\right)^2\right]\rho'^2 \tag{7-27}$$

H_{ro} is the energy of the roton in the absence of the phonon, and, ρ' the density operator of the phonon, Eq. 7-3. In performing the expansion, Eq. 7-27, we neglect terms containing the difference $p - p_0$ since most of the rotons have momenta p close to p_0. We shall likewise discard in what follows, the term in Eq. 7-27 containing the derivative $\partial\Delta/\partial\rho$ since it is obviously smaller than the the term 7-26. The phonon-roton interaction energy can therefore finally be written in the form

$$V = -\tfrac{1}{2}(\mathbf{p}\cdot\mathbf{v} + \mathbf{v}\cdot\mathbf{p}) + \frac{1}{2}\left[\frac{\partial^2\Delta}{\partial\rho^2} + \frac{1}{\mu}\left(\frac{\partial p_0}{\partial\rho}\right)^2\right]\rho'^2 \tag{7-28}$$

The phonon-roton scattering process turns out to be a two-phonon process, since the transitions that interest us are obtained in second-order perturbation theory from the linear term in v in Eq. 7-28 and in first order from the terms quadratic in ρ'.†

†The fact that the equations of hydrodynamics are linear complicates the picture slightly. Indeed, if we solve these equations by successive approximations, then from a superposition of the first-order plane-wave solutions we get in second-order terms containing products of phonon amplitudes. Consequently, already in the first approximation of perturbation theory, the first

If we take into account the fact that the phonon momentum p is small compared to the roton momentum p_0 and that the phonon energy is small compared with μc^2, the transition amplitude for the process under consideration has the following form:

$$\langle n | H | n' \rangle = \frac{p_0 p}{2\rho_0} \left\{ (\mathbf{n} + \mathbf{n}') \cdot \mathbf{m}\, (\mathbf{n} \cdot \mathbf{n}') + \frac{p_0}{\mu c} (\mathbf{n} \cdot \mathbf{m})^2 (\mathbf{n}' \cdot \mathbf{m})^2 + A^2 \right\} \tag{7-29}$$

$$A = \frac{\rho_0^2}{p_0 c} \left[\frac{\partial^2 \Delta}{\partial \rho^2} + \frac{1}{\mu} \left(\frac{\partial p_0}{\partial \rho} \right)^2 \right]$$

Here $\mathbf{n}$ and $\mathbf{n}'$ are unit vectors in the direction of the incoming and outgoing phonons, respectively, and $\mathbf{m}$, a unit vector in the direction of the roton momentum.

From the standpoint of conservation laws, phonon-roton scattering is similar to the scattering of a light particle by a heavy one. The magnitude of the phonon momentum and the direction of the roton momentum are left practically unchanged by the scattering process. With the help of Eq. 7-29 we can find the differential cross section

$$d\sigma = \left(\frac{p_0 p^2}{4\pi \hbar^2 \rho_0 c} \right)^2 \left\{ (\mathbf{n} + \mathbf{n}') \cdot \mathbf{m}\, (\mathbf{n} \cdot \mathbf{n}') \right.$$

$$\left. + \frac{p_0}{\mu c} (\mathbf{m} \cdot \mathbf{n})^2 (\mathbf{m} \cdot \mathbf{n}')^2 + A^2 \right\} do' \tag{7-30}$$

Upon averaging this expression over all directions of the roton momentum we finally obtain

$$d\sigma = \left(\frac{p_0 p^2}{4\pi \hbar^2 \rho_0 c} \right)^2 \left\{ \frac{2}{3} (1 + \cos\psi) \cos^2 \psi \right.$$

$$+ \frac{1}{105} \left(\frac{p_0}{\mu c} \right) \left(1 + 8 \cos^2 \psi + \frac{8}{3} \cos^4 \psi \right)$$

$$\left. + \frac{2A}{15} \left(\frac{p_0}{\mu c} \right) (1 + 2 \cos^2 \psi) + A^2 \right\} do' \tag{7-31}$$

---

term in Eq. 7-28 can lead to the required transitions. A more detailed analysis, however, shows that in second order, the velocity $\mathbf{v}$ contains products of amplitudes with factors $\mathbf{p} - \mathbf{p}'$, where $\mathbf{p}$ and $\mathbf{p}'$ are the initial and final momenta of the phonon. Since the momentum of the phonon is much smaller than the momentum of a roton $\mathbf{p}_0$, the scattering is elastic due to the conservation laws, and $\mathbf{p} \approx \mathbf{p}'$. It follows therefore, that the above-mentioned effect may be neglected.

The angle ψ which enters into this formula is the angle between the incoming and outgoing phonons. The total cross section is obtained by integrating Eq. 7-31 over all scattering angles

$$\sigma_{\mathrm{ph\,r}} = \frac{1}{4\pi}\left(\frac{p_0 p^2}{\hbar^2 \rho_0 c}\right)^2 \left\{\frac{2}{9} + \frac{1}{25}\left(\frac{p_0}{\mu c}\right)^2 + \frac{2A}{9}\frac{p_0}{\mu c} + A^2\right\} \tag{7-32}$$

We give here the approximate values for the derivatives with respect to the density of the parameters of the spectrum, as found from experimental data

$$\partial \ln p_0/\partial \ln \rho \approx 0,4 \quad \partial \ln \Delta/\partial \ln \rho \approx -0,57 \quad (\rho^2/\Delta)(\partial^2\Delta/\partial\rho^2) \approx -5$$

$$A \approx -0,1 \tag{7-33}$$

ROTON-ROTON SCATTERING

The theory gives no information on the character of the interaction between rotons. However in order to calculate the temperature dependence of the characteristic times for roton-roton scattering, it turns out that it is sufficient to know the corresponding probabilities to within a constant factor. Such a probability is quite insensitive to the exact choice of the interaction energy. We may, therefore, take the roton interaction to be a δ-function of the distance between the rotons

$$V = V_0\, \delta(\mathbf{r} - \mathbf{r}_1) \tag{7-34}$$

$\mathbf{r}$ and $\mathbf{r}_1$ being the radius vectors of the rotons and V_0 some constant.

We choose as wave function plane waves symmetrized pairwise over the incoming and outgoing rotons. With these wave functions let us calculate the matrix element for the transition. In order to find the transition probability, dw, we integrate the square of the amplitude of the matrix element over the phase space of one of the scattered rotons

$$dw = \frac{2\pi}{\hbar}\,|2V_0|^2\, \delta(\epsilon + \epsilon_1 - \epsilon' - \epsilon_1')\, \frac{d\mathbf{p}'}{(2\pi\hbar)^2} \tag{7-35}$$

From the form of the energy spectrum it follows that most of the rotons will have momenta whose magnitude is close to p_0. Consequently the changes in roton momenta due to scattering will be significantly smaller than p_0 in magnitude.

Let the angle between the roton momenta $\mathbf{p}$ and $\mathbf{p}_1$ before scattering be θ. Then from Fig. 3 it is easy to see that after the collision the roton momenta can be written in the form

$$p' = p_0 + f_x \cos\frac{\theta}{2} + f_y \sin\frac{\theta}{2}$$
$$p_1' = p_0 + f_x \cos\frac{\theta}{2} - f_y \sin\frac{\theta}{2} \tag{7-36}$$

where we have introduced the variable f, and we have $|f| \ll p_0$. With the new variables the phase space volume element takes the form

$$dp' = 2\pi p_0 \sin\frac{\theta}{2}\, df_x\, df_y \tag{7-37}$$

and the law of conservation of energy the form

$$f_x^2 \cos^2\frac{\theta}{2} + f_y^2 \sin^2\frac{\theta}{2} = \tfrac{1}{2}(p - p_0)^2 + \tfrac{1}{2}(p_1 - p_0)^2 \tag{7-38}$$

In order to calculate the total scattering cross section it is necessary to integrate Eq. 7-36 over the phase space of the scattered particle. The only dependence on the coordinates of phase space which occurs in Eq. 7-36 is in the δ-function which expresses the law of energy conservation.

If, for convenience in the integration, we introduce the auxiliary variable g defined by

$$g^2 = f_x^2 \cos^2\frac{\theta}{2} + f_y^2 \sin^2\frac{\theta}{2}$$

then we may use Eqs. 7-37 and 7-38 to perform the integration of the δ-function over the phase space of the scattered particle, obtaining

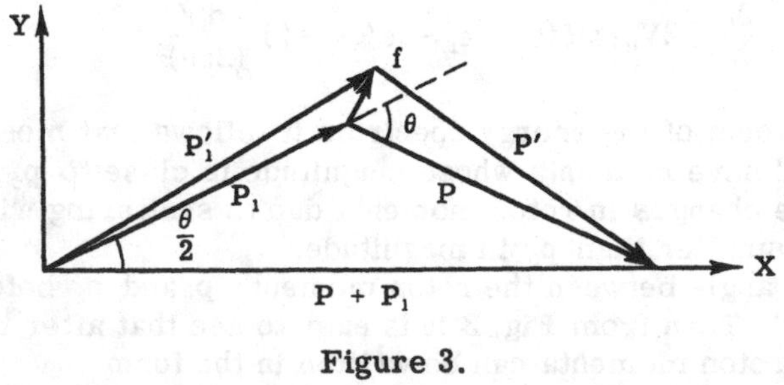

Figure 3.

$$w = \frac{2p_0 \mu |V_0|^2}{\hbar^4 \cos^2 \frac{\theta}{2}} \tag{7-39}$$

The inverse of the time t_r between two collisions can be obtained from Eq. 7-39 by multiplying by $\mathfrak{N}_r$ the roton density defined by Eq. 2-8′, and by averaging over all angles between the colliding rotons. We thus get

$$\frac{1}{t_r} = \frac{4p_0 \mu |V_0|^2 \mathfrak{N}_r}{\hbar^4} \tag{7-40}$$

The constant $|V_0|^2$ is approximately equal to 3.5×10^{-76}, as estimated by using data on the viscosity of helium.

THE ABSORPTION AND EMISSION OF ROTONS AND PHONONS

When elementary excitations collide there can occur inelastic processes in which the number of quasi-particles changes. These processes may be divided into three types:

1. Processes of emission (or absorption) of phonons.
2. Processes of transformation of rotons into phonons and vice versa
3. Processes of emission (or absorption) of rotons.

We shall only be interested in the most probable processes of each type.

The most probable process of type (1) is the five-phonon process in which two phonons are transformed into three. The probability for such a process can be calculated by perturbation theory analogously to the four-phonon process. The transition amplitude that interests us is obtained in third-order perturbation theory from the anharmonic terms V_3 in Eq. 7-13, in second order from the terms V_4 and V_3, and in first order from terms V_5 in expansion 7-12. In performing the sums over intermediate states we again find vanishing denominators when the momenta of the incident and intermediate phonons have nearly the same direction. Therefore the third-order perturbation theory terms will give the largest contributions since they each contain two vanishing denominators. One must, of course, take into account the dispersion in the phonon spectrum, since otherwise the angular integrations would diverge. We shall not perform here the full calculation of the probability of the five-phonon process, since the answer contains the square of the dispersion parameter γ, which is poorly known.† In order to determine merely the temperature dependence of the appropriate kinetic coefficients, it is sufficient to know the dependence of the probabilities on the phonon momenta. Simple qualitative arguments permit one to conclude that the angular average of the probability of the five-phonon process is proportional to the cube of some momentum.

$$\overline{w} \sim p^3 \tag{7-41}$$

In what follows it turns out that the precise knowledge of which momenta enter into w is not necessary. Let the total number of phonons per unit volume be $\mathfrak{N}_{ph}$ (in general this is not equal to the equilibrium number). The rate of change of the number of phonons due to five-phonon processes can be written in the form

$$\dot{\mathfrak{N}}_{ph} = -\iiiint \{n_1 n_2 n_3 (n_4 + 1)(n_5 + 1) - (n_1 + 1)(n_2 + 1) \times (n_3 + 1) n_4 n_5\} \cdot dw \frac{dp_1\, dp_2\, dp_3}{(2\pi\hbar)^9} \tag{7-42}$$

If the total number of phonons is not equal to its equilibrium value this means that the distribution functions n contain a nonzero chemical potential μ_{ph},

$$n = \{\exp[(\epsilon - \mu_{ph})/kT] - 1\}^{-1}$$

For small deviations from equilibrium we can expand the function n in powers of μ_{ph} and limit ourselves to linear terms; this yields

$$n - n_0 = n_0(n_0 + 1)\mu_{ph}/kT \tag{7-43}$$

The function n_0 is the equilibrium phonon distribution function, corresponding to $\mu_{ph} = 0$. By performing some simple transformations using Eq. 7-43 we may write Eq. 7-42 in the form

$$\dot{\mathfrak{N}}_{ph} = -\iiiint n_{10} n_{20} n_{30} (n_{40} + 1)(n_{50} + 1)\, dw \frac{dp_1\, dp_2\, dp_3}{(2\pi\hbar)^9} \frac{\mu_{ph}}{kT}$$

---

†According to the neutronographic data, obtained after this analysis was performed, it has been found that the parameter is negative. It means that the phonon part of the spectrum is unstable, i.e., the decay of a phonon into two phonons is permitted by conservation laws (three-phonon processes). The account of three-phonon processes does not qualitatively affect the overall picture of dissipation effects.

Let us denote by Γ_{ph} the quantity relating $\mathfrak{N}_{ph}$ and μ_{ph}. We have

$$\Gamma_{ph} = \frac{1}{kT} \iiiint n_{10} n_{20} n_{30} (n_{40} + 1)(n_{50} = 1) \, dw \, \frac{dp_1 \, dp_2 \, dp_3}{(2\pi\hbar)^9} \tag{7-44}$$

We may neglect the distribution functions n_{50} and n_{40} in comparison with unity in Eq. 7-44 without destroying significantly the accuracy of our calculation. We may then perform the integrals over the phase space of the three colliding phonons independently ($\mathbf{p}_1$, $\mathbf{p}_2$, and $\mathbf{p}_3$); this leads to the replacement of dw by $\overline{w}$ and reduces the number of integrations by one. We thus obtain

$$\Gamma_{ph} \approx \frac{1}{kT} \iiint n_{10} n_{20} n_{30} \overline{w} \, \frac{dp_1 \, dp_2 \, dp_3}{(2\pi\hbar)^9} \tag{7-44'}$$

The integrand in Eq. 7-44′ is proportional to p^{12}. We may now integrate Eq. 7-44′ over the phase space of the colliding phonons to find the temperature dependence of Γ_{ph}:

$$\Gamma_{ph} = aT^{11} \tag{7-45}$$

where a is a temperature-independent coefficient.

Relation 7-45 determines the temperature dependence of the quantity that will be important in what follows. By analyzing experiments on the absorption of sound we can determine the approximate value $a \sim 1 \times 10^{43}$ for the coefficient.

Let us now consider case (b). A phonon cannot decay directly into a roton since its momentum p is much less than the roton momentum p_0. When an energetic phonon (whose energy is of order Δ) collides with a roton, two rotons may be formed. From the momentum conservation law we see that the angle formed by the created rotons may not be too small. As in the previous case, we may estimate the probability for such a process. To do this, we can consider that the high-energy phonon interacts with the roton in the same way as a roton would—that is, that the phonon-roton interaction has the δ-function form of Eq. 7-34. Just as before, the rate of approach of the phonon and roton numbers to their equilibrium values can be expressed in terms of the corresponding chemical potentials

$$\dot{\mathfrak{N}}_r = \Gamma_{ph\,r}(\mu_r - \mu_{ph}) \qquad \dot{\mathfrak{N}}_{ph} = \Gamma_{ph\,r}(\mu_r - \mu_{ph}) \tag{7-46}$$

The coefficient $\Gamma_{ph\,r}$ is determined from the total collision integral for the process under consideration, namely the decay of a phonon into a roton (and vice versa). We shall not perform the intermediate calculations here, but merely write down the final result

$$\Gamma_{ph\,r} = |V_0|^2 \Delta^2 \mathfrak{N}_r^2 / \pi \hbar^4 c^3 kT \tag{7-47}$$

The amplitude V_0 for this process is not known; however we may obtain an order of magnitude estimate for the quantity $\Gamma_{ph\,r}$ by taking for V_0 the corresponding quantity in the case of roton-roton interactions. For the purposes of our further investigations of kinetic phonomena, we separate out of $\Gamma_{ph\,r}$ its temperature dependence and write Eq. 7-47 in the form

$$\Gamma_{ph\,r} = b\, e^{-2\Delta/T} \tag{7-48}$$

The coefficient b can again be determined by analyzing the corresponding experiments on the absorption of sound; one then finds approximately $b \sim 4 \times 10^{50}$.

As for process (c), namely, the transformation of two rotons into three (and vice versa), it turns out to be extremely unlikely, due to strong energy and momentum restrictions, and does not play any role in the phenomena that interest us.

PART II

HYDRODYNAMICS

8

THE HYDRODYNAMICS OF SUPERFLUIDS[12]

By using the microscopic description of a superfluid which was outlined in the preceding sections, one can construct the complete system of hydrodynamic equations valid in this case. The fundamental assumptions upon which our whole analysis will be based are the following: The ordered motion of the excitations carries along with it only part of the liquid, characterized by the "normal" density ρ_n. The remaining part, the "superfluid," is characterized by the density $\rho_s = \rho - \rho_n$, and performs an independent motion. This latter motion has the important property of being irrotational. Thus in a superfluid, there can exist two simultaneous but independent motions, one normal and the other superfluid, with velocities $\mathbf{v}_n$ and $\mathbf{v}_s$, with

$$\text{curl } \mathbf{v}_s = 0 \tag{8-1}$$

Condition 8-1 which states that superfluid motion is irrotational will not be violated until the flow velocities have reached certain critical values where the normal and superfluid parts of the liquid begin to interact.

We may now proceed to the derivation of the complete system of hydrodynamic equations, taking as our starting point the conservation laws and the Galilean relativity principle.

The conservation laws have the same form for all physical quantities; they are differential equations stating that the time derivative of the conserved quantity is equal to the divergence of some vector.

The law of conservation of mass is the so-called continuity equation, which relates the total density ρ and the current $\mathbf{j}$ (the momentum per unit volume)

$$\dot{\rho} + \operatorname{div} \mathbf{j} = 0 \tag{8-2}$$

The momentum conservation law is the equation of motion

$$\frac{\partial}{\partial t} j_i + \frac{\partial \Pi_{ik}}{\partial r_k} = 0 \tag{8-3}$$

where Π_{ik} is the momentum flux density tensor.

We shall not at present consider dissipative processes. Then the flow is reversible and the entropy S is also conserved. We may therefore write

$$\dot{S} + \operatorname{div} \mathbf{F} = 0 \tag{8-4}$$

where $\mathbf{F}$ is the entropy flux vector. Since the entropy is only connected to the excitations it should be carried along with the normal motion. Consequently the entropy flux is equal to $S\mathbf{v}_n$. However we shall not at this point make any reference to the microscopic picture; later on, we shall show that the relation $\mathbf{F} = S\mathbf{v}_n$ follows from the conservation laws. For the moment, therefore, $\mathbf{F}$ remains some unknown vector to be determined.

In a superfluid there can be two different motions and consequently there should be two hydrodynamic equations of motion. The first one is Eq. 8-3 and the second one an equation for the time derivative of $\mathbf{v}_s$. Since $\operatorname{curl} \mathbf{v}_s = 0$, we can write

$$\dot{\mathbf{v}}_s + \nabla\left(\varphi + \frac{\mathbf{v}_s^2}{2}\right) = 0 \tag{8-5}$$

where φ is some scalar function.

Equations 8-2 to 8-5 represent the complete system of hydrodynamic equations of a superfluid. In order to make them completely meaningful we must still determine the form of the unknown terms Π_{ik}, $\mathbf{F}$, and φ. In order to do this we use the energy conservation law which can be written in differential form

$$\frac{\partial E}{\partial t} + \operatorname{div} \mathbf{Q} = 0 \tag{8-6}$$

where E is the energy per unit volume of liquid, and $\mathbf{Q}$ the energy flux. It is necessary to choose the unknown terms in Eqs. 8-2 to 8-5 in such a way that Eq. 8-6 be automatically satisfied. We may further use the Galilean relativity principle which allows us to determine the dependence of all quantities on the velocity $\mathbf{v}_s$ for a given value of the difference $\mathbf{v}_n - \mathbf{v}_s$.

In what follows it will be useful to consider a new frame of reference (K_0) in which the velocity of superfluid motion of a given element of the liquid is equal to zero. The frame K_0 moves at a velocity $\mathbf{v}_s$ with respect to the original frame K. The values of the quantities that interest us in the different frames are related by the well-known formulas†

$$\mathbf{j} = \rho \mathbf{v}_s + \mathbf{j}_0 \tag{8-7}$$

$$\Pi_{ik} = \rho v_{si} v_{sk} + v_{si} j_{0k} + v_{sk} j_{0i} + \pi_{ik} \tag{8-8}$$

$$E = \frac{\rho v_s^2}{2} + \mathbf{v}_s \cdot \mathbf{j}_0 + E_0 \tag{8-9}$$

$$\mathbf{Q} = \left(\frac{\rho v_s^2}{2} + \mathbf{v}_s \cdot \mathbf{j}_0 + E_0\right)\mathbf{v}_s + \frac{v_s^2}{2}\mathbf{j}_0 + \pi \cdot \mathbf{v}_s + \mathbf{q} \tag{8-10}$$

$$\mathbf{F} = S\mathbf{v}_s + \mathbf{f} \tag{8-11}$$

Here $\mathbf{j}_0$ is the momentum, π_{ik} the momentum flux tensor, E_0 the energy, $\mathbf{q}$ the energy flux vector, and $\mathbf{f}$ the entropy flux in the frame K_0.

In the frame K_0 the liquid moves with the velocity $\mathbf{v}_n - \mathbf{v}_s$ and obviously all quantities ($\mathbf{j}_0$, π_{ik}, E_0, $\mathbf{q}$, $\mathbf{f}$) may only depend on this difference. The energy E_0 satisfies the thermodynamic identity

---

†Formulas 8-7 to 8-11 follow immediately from the Galilean principle of relativity. Let us show, for instance, how relation 8-8 may be proved for an ordinary liquid. The momentum flux tensor in hydrodynamics is equal to

$$\Pi_{ik} = \rho u_i u_k + p\,\delta_{ik}$$

where u is the velocity and p the pressure. The velocity of the liquid in the rest frame is related to its value u′ in a frame moving with velocity $\mathbf{v}$ by the formula

$$\mathbf{u} = \mathbf{u}' + \mathbf{v}$$

Inserting u′ in the expression for Π_{ik} yields

$$\Pi_{ik} = \rho v_i v_k + \rho u_i' v_k + \rho u_k' v_i + (\rho u_i' u_k' + p\,\delta_{ik})$$

If we denote by $\mathbf{j}_0$ the momentum per unit volume in the moving frame, we obtain the desired relation

$$\Pi_{ik} = \rho v_i v_k + v_i j_{0k} + v_k j_{0i} + \pi_{ik}$$

This formula retains the same form in superfluid hydrodynamics. The other formulas may be proved analogously.

$$dE_0 = T\,dS + \mu\,d\rho + (\mathbf{v}_n - \mathbf{v}_s)\cdot d\mathbf{j}_0 \tag{8-12}$$

Here μ is the chemical potential, T the temperature. The third term in Eq. 8-12 simply expresses the fact that the velocity is the derivative of the energy with respect to the momentum and should be looked upon as a definition of the velocity $\mathbf{v}_n$. By symmetry considerations it follows that the vector $\mathbf{j}_0$ may only be directed along $\mathbf{v}_n - \mathbf{v}_s$, so that we may write

$$\mathbf{j}_0 = \rho_n(\mathbf{v}_n - \mathbf{v}_s) \tag{8-13}$$

Relation 8-13 should be considered as the definition of the normal density ρ_n. From Eqs. 8-7 and 8-13 we then find

$$\mathbf{j} = \rho_s \mathbf{v}_s + \rho_n \mathbf{v}_n \tag{8-14}$$

where $\rho_s = \rho - \rho_n$.

The ensuing calculation may be summed up as follows: we differentiate the energy E with respect to time, and express all time derivatives of thermodynamic quantities, namely of $\mathbf{j}$ and $\mathbf{v}_s$, by means of Eqs. 8-2 to 8-5. We then calculate div $\mathbf{Q}$ using Eq. 8-10 and insert $\dot{E}$ and div $\mathbf{Q}$ into Eq. 8-6. After a considerable amount of rearrangement we obtain

$$\begin{aligned} \operatorname{div}\mathbf{q} = {} & -(\mathrm{m}\cdot\nabla)\cdot\mathbf{v}_s + (\mathbf{v}_n - \mathbf{v}_s)\cdot(\nabla\cdot\mathrm{m}) + \mathbf{j}_0\cdot((\mathbf{v}_n - \mathbf{v}_s)\cdot\nabla)\mathbf{v}_n \\ & + (\mathbf{j}_0 - \rho(\mathbf{v}_n - \mathbf{v}_s))\cdot\nabla(\varphi - \mu) - \nabla T\cdot(\mathbf{f} - S(\mathbf{v}_n - \mathbf{v}_s)) \\ & + \operatorname{div}(\mathbf{f}T + \mathbf{j}_0\mu) \end{aligned} \tag{8-15}$$

Instead of π_{ik} it is convenient to introduce another tensor

$$m_{ik} = \pi_{ik} + [E_0 - TS - \mu\rho - ((\mathbf{v}_n - \mathbf{v}_s)\cdot\mathbf{j}_0)]\,\delta_{ik} \tag{8-16}$$

In the absence of energy dissipation the quantities m_{ik}, $\mathbf{q}$, $\mathbf{f}$, and φ are functions of the thermodynamic variables and the velocities, but do not depend on their space or time derivatives. This enables us to obtain from Eq. 8-15 unique expressions for the desired quantities

$$m_{ik} = j_{oi}(v_{nk} - v_{sk}) \qquad \mathbf{f} = S(\mathbf{v}_n - \mathbf{v}_s) \qquad \varphi = \mu$$

$$\mathbf{q} = T\mathbf{f} + \mu\mathbf{j}_0 - ((\mathbf{v}_n - \mathbf{v}_s)\cdot\mathbf{v}_s)\,\mathbf{j}_0 + (\mathbf{v}_n - \mathbf{v}_s)(\mathbf{v}_n\cdot\mathbf{j}_0) \tag{8-17}$$

Finally we may write

$$\mathbf{F} = \mathbf{f} + S\mathbf{v}_s = S\mathbf{v}_n \tag{8-18}$$

$$\Pi_{ik} = \rho v_{si} v_{sk} + v_{si} j_{ok} + v_{nk} j_{oi} - [E_0 - TS - \mu\rho - ((\mathbf{v}_n - \mathbf{v}_s)\cdot \mathbf{j}_0)]\,\delta_{ik} \tag{8-19}$$

$$\mathbf{Q} = \left(\mu + \frac{\mathbf{v}_s^2}{2}\right)(\mathbf{j}_0 + \rho\mathbf{v}_s) + ST\mathbf{v}_n + \mathbf{v}_n(\mathbf{v}_n \cdot \mathbf{j}_0) \tag{8-20}$$

The expression in square brackets in Eq. 8-19 is the pressure; by definition it is equal to the derivative of the total energy with respect to the volume, when the total mass, the total entropy, and the total momentum of relative motion are all held constant.

$$p = \frac{\partial(E_0 V)}{\partial V} = -E_0 + TS + \mu\rho + ((\mathbf{v}_n - \mathbf{v}_s)\cdot \mathbf{j}_0) \tag{8-21}$$

Inserting the expressions for $\mathbf{F}$, φ, and Π_{ik} into Eqs. 8-3 to 8-5 we obtain the complete system of hydrodynamic equations for superfluids

$$\dot{\rho} + \operatorname{div} \mathbf{j} = 0 \tag{8-22}$$

$$\dot{\mathbf{j}} + \mathbf{v}_s \operatorname{div} \mathbf{j} + (\mathbf{j}\cdot\nabla)\mathbf{v}_s + \mathbf{j}_0 \operatorname{div} \mathbf{v}_n + (\mathbf{v}_n\cdot\nabla)\mathbf{j}_0 + \nabla p = 0 \tag{8-23}$$

$$\dot{S} + \operatorname{div} S\mathbf{v}_n = 0 \tag{8-24}$$

$$\dot{\mathbf{v}}_s + \nabla\left(\frac{\mathbf{v}_s^2}{2} + \mu\right) = 0 \tag{8-25}$$

The momentum flux tensor Π_{ik} may, by Eq. 8-14, be put into the form

$$\Pi_{ik} = \rho v_{ni} v_{nk} + \rho_s v_{si} \cdot v_{sk} + p\,\delta_{ik} \tag{8-19'}$$

where the first term is the momentum flux of the normal motion, and the second term the momentum flux of the superfluid motion. The hydrodynamic equations we have obtained in Eqs. 8-22 to 8-25 are rather complicated since the quantities μ, ρ_n, S, etc., which appear in them, are functions of the relative velocity $\mathbf{v}_n - \mathbf{v}_s$, whose form can only be determined by going to the microscopic theory.

These general hydrodynamic equations are considerably simplified in the case of small velocities. It must be noted that the property of superfluidity is destroyed when the velocities are greater than certain critical values. However under nonstationary conditions, for instance for sound propagation, the velocities may become considerably larger than these critical values. There exists therefore a region of applicability of the general equations where their nonlinear character becomes apparent (cf. Chap. 13). If we restrict ourselves to quadratic terms in the velocities we may neglect the dependence of ρ_n and ρ_s on the velocities. Let us choose as our independent thermodynamic variables the pressure p and the temperature T. Let us write the thermodynamic identity satisfied by the chemical potential, which by Eqs. 8-12 and 8-21 may be written

$$d\mu = -\sigma\, dT + \frac{1}{\rho}\, dp - \frac{\rho_n}{\rho}(\mathbf{v}_n - \mathbf{v}_s)\cdot d(\mathbf{v}_n - \mathbf{v}_s) \qquad \left(\sigma = \frac{S}{\rho}\right) \tag{8-26}$$

From this relation it is easy to find the dependence of the entropy σ and the density ρ on the relative velocity $\mathbf{w} = \mathbf{v}_n - \mathbf{v}_s$. From the general relations for derivatives

$$\frac{\partial \sigma}{\partial (w^2)} = \frac{1}{2}\frac{\partial}{\partial T}\frac{\rho_n}{\rho} \qquad \frac{\partial \rho}{\partial (w^2)} = \rho^2 \frac{1}{2}\frac{\partial}{\partial p}\frac{\rho_n}{\rho} \tag{8-27}$$

we can find the first terms in the expansion of σ and ρ in powers of w^2

$$\sigma(p,T,w) = \sigma(p,T) + \tfrac{1}{2}w^2 \frac{\partial}{\partial T}\frac{\rho_n}{\rho} \tag{8-28}$$

$$\rho(p,T,w) = \rho(p,T) + \tfrac{1}{2}\rho^2 w^2 \frac{\partial}{\partial p}\frac{\rho_n}{\rho} \tag{8-29}$$

Inserting these expressions into the general equations, 8-22 to 8-25, we obtain equations that are valid up to second order in the velocities

$$\frac{\partial}{\partial t}\left(\rho + \tfrac{1}{2}w^2\rho^2 \frac{\partial}{\partial p}\frac{\rho_n}{\rho}\right) + \operatorname{div}\mathbf{j} = 0 \qquad \mathbf{j} = \rho_s \mathbf{v}_s + \rho_n \mathbf{v}_n \tag{8-30}$$

$$\frac{\partial j_i}{\partial t} + \frac{\partial \Pi_{ik}}{\partial r_k} = 0 \qquad \Pi_{ik} = \rho_s v_{si} v_{sk} + \rho_n v_{ni} v_{nk} + p\,\delta_{ik} \tag{8-31}$$

$$\frac{\partial}{\partial t}\left[S + \frac{\rho w^2}{2}\left(\frac{\partial}{\partial T}\frac{\rho_n}{\rho} + S\frac{\partial}{\partial p}\frac{\rho_n}{\rho}\right)\right] + \operatorname{div} S\mathbf{v}_n = 0 \tag{8-32}$$

$$\frac{\partial \mathbf{v}_s}{\partial t} + \nabla\left[\mu + \frac{v_s^2}{2} - \frac{\rho_n w^2}{2\rho}\right] = 0 \tag{8-33}$$

In formula 8-33 there appears the chemical potential μ of the liquid at rest. In the above approximate equations the velocities are considered small with respect to the velocities of first and second sound. The general equations may also be considerably simplified in the case where the velocities are small only in comparison with the velocity of first sound, but are comparable to the velocity of second sound. In this last case it is possible to find a general solution in the form of a one-dimensional running wave. The considerable simplification is due to the fact that the normal density ρ_n depends only weakly on the square of the relative velocity $\mathbf{w}^2$.

Let us now consider the question of the boundary conditions satisfied by the thermodynamic quantities. Obviously the normal component of the current j should vanish at the walls, since there can be no transport of matter across this boundary. As for the velocity of normal motion $\mathbf{v}_n$, it is linked to the motion of the gas of excitations, which has all the properties of a normal viscous liquid. Consequently the tangential component of $\mathbf{v}_n$ should vanish at the boundary of a solid body. The normal component of $\mathbf{v}_n$ (directed along the z-axis) is not equal to zero, but determines the heat flux from the liquid to the solid body, which according to Eq. 8-20, for $j_z = 0$, is equal to STv_{nz}. The normal component of the heat flux is continuous across the boundary between the liquid and the solid body. The temperature has a discontinuity across the boundary (cf. Chap. 23), which is proportional to the magnitude of the heat flux.

Let us choose the x and y axes along the surface of the solid and the z-axis perpendicular to it. Then the above boundary conditions are written as follows

$$\rho_s v_{sz} + \rho_n v_{nz} = 0 \qquad v_{nx} = v_{ny} = 0 \tag{8-34}$$

$$STv_{nz} = -\kappa_s \left(\frac{\partial T}{\partial z}\right)_s \qquad T_\ell - T_s = Kq_z \tag{8-35}$$

Here κ_s is the thermal conductivity of the solid.

In many cases it is possible to neglect the thermal conductivity of the solid and to set κ_s equal to zero. This gives

$$v_{nz} = 0 \qquad v_{sz} = 0 \tag{8-36}$$

that is, the boundary conditions for $\mathbf{v}_n$ are those of a normal liquid and for $\mathbf{v}_s$ those of an ideal liquid.

A moving superfluid in the presence of a current normal to the walls causes tangential forces to act on the surface of a solid body. This can be seen from the fact that the component Π_{xz} of the momentum flux tensor is nonzero in this case. Indeed, from the first of relations 8-34 we can find this component

$$\Pi_{xz} = \rho_s v_{sx} v_{sz} + \rho_n v_{nx} v_{nz} = \rho_n v_{nz} (v_{nx} - v_{sx})$$

Expressing the component v_{nz} in terms of the normal component of the heat flux q_z we finally obtain

$$\Pi_{xz} = \frac{\rho_n q_z}{ST} (v_{nx} - v_{sx}) \qquad (8\text{-}37)$$

9

THE DISSIPATIVE TERMS IN THE HYDRODYNAMIC EQUATIONS[13]

The system of equations, 8-22 to 8-25, obtained above describes the motion of a superfluid in the absence of energy dissipation. In reality, the lack of equilibrium leads to the appearance in all the fluxes, of terms depending on the derivatives of the velocities and thermodynamic variables, with respect to the coordinates. It must be noted that under nonequilibrium conditions the usual definitions of thermodynamic quantities lose their meaning and must be made more precise. If as before we denote by ρ the mass per unit volume of liquid and by $\mathbf{j}$ its momentum, then the continuity equation retains its usual form

$$\rho + \operatorname{div} \mathbf{j} = 0 \tag{9-1}$$

We shall further denote by E_0 the energy per unit volume in the frame of reference in which the superfluid part is at rest. The remaining thermodynamic variables are defined to have the same functional dependence on the density ρ, the energy E, and the relative velocity w as they do in thermodynamic equilibrium. Here the entropy $S(\rho,E,w)$ will not be the true entropy, whose integral $\int S\, dV$ must always increase with time. However for situations close to equilibrium, when the gradients of all quantities are small, the entropy defined in this manner will be almost identical to the true entropy. Indeed it is easy to see that there can be no terms linear in the gradients in the power expansion of the entropy, since these are odd. The entropy attains its maximum value in the equilibrium state; consequently its expansion in powers of the gradients begins with quadratic terms, which may be neglected in the approximation we are considering.

The law of conservation of momentum can, as previously, be written in the form

$$\dot{j}_i + \frac{\partial}{\partial r_k}(\Pi_{ik} + \tau_{ik}) = 0 \qquad (9\text{-}2)$$

with the difference that in the momentum flux, along with the usual terms, there occurs an unknown dissipative term τ_{ik}, which must still be determined. In a similar manner we may add another term ∇h in the equation of superfluid flow, Eq. 8-25,

$$\dot{\mathbf{v}}_s + \nabla\left(\mu + \frac{\mathbf{v}_s^2}{2} + h\right) = 0 \qquad (9\text{-}3)$$

(but as before curl $\mathbf{v}_s = 0$!).

Obviously the expression for the heat flux in the law of conservation of energy is also changed by some quantity $\mathbf{Q}'$

$$\frac{\partial E}{\partial t} + \operatorname{div}(\mathbf{Q} + \mathbf{Q}') = 0 \qquad (9\text{-}4)$$

As for the entropy equation, it does not now have the form of a continuity equation, since entropy is not conserved, but increases. We shall use the requirement that the entropy should increase, in order to determine the unknown dissipative coefficients.

Let us differentiate expression (8-9) for E with respect to time; this yields

$$\frac{\partial E}{\partial t} = \left(\frac{\mathbf{v}_s^2}{2} + \mu\right)\dot{\rho} + \dot{\mathbf{v}}_s \cdot \mathbf{j} + \mathbf{v}_n \cdot \dot{\mathbf{j}}_0 + T\dot{S} \qquad (9\text{-}5)$$

We may then eliminate the time derivatives by Eqs. 9-2 to 9-3, obtaining

$$\frac{\partial E}{\partial t} = \operatorname{div}\{\mathbf{Q} + \mathbf{q} + h(\mathbf{j} - \rho\mathbf{v}_n) + (\tau \cdot \mathbf{v}_n)\}$$

$$= T\left(\dot{S} + \operatorname{div}\left(S\mathbf{v}_n + \frac{\mathbf{q}}{T}\right)\right) + h\operatorname{div}(\mathbf{j} - \rho\mathbf{v}_n)$$

$$+ \tau_{ik}\frac{\partial v_{ni}}{\partial r_k} + \frac{1}{T}\mathbf{q}\cdot\nabla T \qquad (9\text{-}6)$$

Since in our nonequilibrium situation we may also have an additional term in the entropy flux, we here added to both sides of Eq. 9-6 a term, div $\mathbf{q}$, containing the unknown quantity $\mathbf{q}$.

Comparing Eqs. 9-6 and 9-4 we obtain the equation that determines the rate of increase of entropy

$$T\left(\frac{\partial S}{\partial t} + \operatorname{div}\left(S\mathbf{v}_n + \frac{\mathbf{q}}{T}\right)\right) = -h \operatorname{div}(\mathbf{j} - \rho\mathbf{v}_n) - \tau_{ik}\frac{\partial v_{ni}}{\partial r_k} - \frac{1}{T}\mathbf{q}\cdot\nabla T \tag{9-7}$$

and the expression for the additional dissipative flux of heat

$$\mathbf{Q}' = \mathbf{q} + h(\mathbf{j} - \rho\mathbf{v}_n) + (\tau\cdot\mathbf{v}_n) \tag{9-8}$$

The expression on the right-hand side of Eq. 9-7 is the dissipative function of the superfluid. If the spatial derivatives of the velocities and thermodynamic variables are not too large, then in the first approximation all additions to the equations (τ_{ik}, h, $\mathbf{q}$) are linear functions of these derivatives. From the law of increase of entropy it then follows that the dissipative function must be a positive definite quadratic form in these same derivatives. From this requirement we may immediately determine the form of the unknown terms

$$\tau_{ik} = -\eta\left(\frac{\partial v_{ni}}{\partial r_k} + \frac{\partial v_{nk}}{\partial r_i} - \frac{2}{3}\delta_{ik}\frac{\partial v_{n\ell}}{\partial r_\ell}\right) - \delta_{ik}[\zeta_1 \operatorname{div}(\mathbf{j} - \rho\mathbf{v}_n) + \zeta_2 \operatorname{div}\mathbf{v}_n] \tag{9-9}$$

$$h = -\zeta_3 \operatorname{div}(\mathbf{j} - \rho\mathbf{v}_n) - \zeta_4 \operatorname{div}\mathbf{v}_n \tag{9-10}$$

$$q = -\kappa\nabla T \tag{9-11}$$

As usual, in the momentum flux τ_{ik} we separate the combination of derivatives of $\mathbf{v}_n$ that has zero trace (first viscosity).

According to the Onsager symmetry principle for kinetic coefficients we have the relation

$$\zeta_1 = \zeta_4 \tag{9-12}$$

The coefficients ζ_1, ζ_2, ζ_3, ζ_4 are the coefficients of second viscosity; there are, therefore, in all, three independent coefficients of second viscosity. η is the coefficient of first viscosity and is entirely connected with the normal motion, and κ is the thermal conduction coefficient. As was to be expected, there is no coefficient analogous to the first viscosity for the superfluid motion.

Let us now write in final form the hydrodynamic equations for a superfluid taking into account the dissipative terms

$$\dot{\rho} + \operatorname{div} \mathbf{j} = 0 \tag{9-13}$$

$$\left.\begin{aligned}\frac{\partial j_i}{\partial t} + \frac{\partial \Pi_{ik}}{\partial r_k} = \frac{\partial}{\partial r_k}\Bigg\{\eta\left(\frac{\partial v_{ni}}{\partial r_k} + \frac{\partial v_{nk}}{\partial r_i} - \frac{2}{3}\,\delta_{ik}\,\frac{\partial v_{n\ell}}{\partial r_\ell}\right)\\ + \delta_{ik}\,\zeta_1 \operatorname{div}(\mathbf{j} - \rho\mathbf{v}_n) + \delta_{ik}\,\zeta_2 \operatorname{div} \mathbf{v}_n\Bigg\}\end{aligned}\right\} \tag{9-14}$$

$$\dot{\mathbf{v}}_s + \nabla\left(\mu + \frac{v_s^2}{2}\right) = \nabla\{\zeta_3 \operatorname{div}(\mathbf{j} - \rho\mathbf{v}_n) + \zeta_4 \operatorname{div} \mathbf{v}_n\} \tag{9-15}$$

$$\dot{S} + \operatorname{div}\left(S\mathbf{v}_n + \frac{\mathbf{q}}{T}\right) = \frac{1}{T}R \tag{9-16}$$

The dissipative function is equal to

$$\begin{aligned}R = \zeta_2(\operatorname{div} \mathbf{v}_n)^2 + \zeta_3(\operatorname{div}(\mathbf{j} - \rho\mathbf{v}_n))^2 + 2\zeta_1 \operatorname{div} \mathbf{v}_n \operatorname{div}(\mathbf{j} - \rho\mathbf{v}_n)\\ + \frac{1}{2}\eta\left(\frac{\partial v_{ni}}{\partial r_k} + \frac{\partial v_{nk}}{\partial r_i} - \frac{2}{3}\,\delta_{ik}\,\frac{\partial v_{n\ell}}{\partial r_\ell}\right)^2 + \kappa\frac{(\nabla T)^2}{T}\end{aligned} \tag{9-17}$$

In order to ensure that the function R will be positive, the kinetic coefficients η, ζ_2, ζ_3, and κ must be positive and ζ_1 must satisfy the inequality

$$\zeta_1^2 \leq \zeta_2 \zeta_3 \tag{9-18}$$

10

THE PROPAGATION OF SOUND IN A SUPERFLUID

In a sound wave the velocities $\mathbf{v}_n$ and $\mathbf{v}_s$ are assumed to be small,† and the thermodynamic quantities almost equal to their equilibrium values. The propagation of sound in helium II is described by the system of hydrodynamic equations, Eqs. 8-22 to 8-25, which in this case may be linearized, yielding

$$\frac{\partial \rho}{\partial t} + \operatorname{div} \mathbf{j} = 0 \tag{10-1}$$

$$\frac{\partial \rho\sigma}{\partial t} + \rho\sigma \operatorname{div} \mathbf{v}_n = 0 \qquad (\sigma\rho = S) \tag{10-2}$$

$$\frac{\partial \mathbf{j}}{\partial t} + \nabla p = 0 \tag{10-3}$$

$$\frac{\partial \mathbf{v}_s}{\partial t} + \nabla \mu = 0 \tag{10-4}$$

Let us eliminate the momentum $\mathbf{j}$ from Eqs. 10-1 and 10-3; we get

$$\frac{\partial^2 \rho}{\partial t^2} = \Delta p \tag{10-5}$$

Furthermore, let us eliminate the velocities $\mathbf{v}_n$ and $\mathbf{v}_s$ from the

†Here we mean that $\mathbf{v}_n$ and $\mathbf{v}_s$ are small compared to the velocity of sound.

three equations, Eqs. 10-2, 10-3, and 10-4. For this, let us take the time derivative of Eq. 10-2 and the divergence of Eqs. 10-3 and 10-4. By eliminating the terms $(\partial/\partial t)\operatorname{div}\mathbf{v}_n$ and $(\partial/\partial t)\operatorname{div}\mathbf{v}_s$ from the equations thus obtained, we find

$$\rho_s\,\Delta\mu - \Delta p + \frac{\rho_n}{\rho\sigma}\frac{\partial^2}{\partial t^2}(\rho\sigma) = 0 \tag{10-6}$$

Let us express the quantity $\partial^2\rho/\partial t^2$ occurring in the above equation by means of Eq. 10-5 and use the thermodynamic identity, Eq. 8-26. We have finally

$$\frac{\partial^2\sigma}{\partial t^2} = \frac{\rho_s}{\rho_n}\sigma^2\,\Delta T \tag{10-7}$$

Equations 10-6 and 10-7 determine the changes in the thermodynamic quantities in the presence of a sound wave.

Let us use, in these equations, the independent variables p and T, which we may represent in the form $p = p_0 + p'$, $T = T_0 + T'$. The quantities with subscript zero denote the equilibrium values, and the primed quantities the deviations from equilibrium due to the sound wave. As a result Eqs. 10-6 and 10-7 take the form

$$\frac{\partial\rho}{\partial p}\frac{\partial^2 p'}{\partial t^2} - \Delta p' + \frac{\partial\rho}{\partial T}\frac{\partial^2 T'}{\partial t^2} = 0 \tag{10-8}$$

$$\frac{\partial\sigma}{\partial p}\frac{\partial^2 p'}{\partial t^2} + \frac{\partial\sigma}{\partial T}\frac{\partial^2 T'}{\partial t^2} - \frac{\sigma^2\rho_s}{\rho_n}\Delta T' = 0 \tag{10-9}$$

Let us look for a solution of the system, Eqs. 10-8 and 10-9, representing a plane wave propagating in some direction. In such a wave the quantities p′ and T′ vary as

$$\exp\left[-i\omega\left(t - \frac{x}{u}\right)\right]$$

(the x-axis can be chosen as the direction of propagation of the wave, ω is the frequency, and u the velocity of sound). The system of equations, Eqs. 10-8 and 10-9, may now be written in the form

$$\left(\frac{\partial\rho}{\partial p}u^2 - 1\right)p' + \frac{\partial\rho}{\partial T}u^2 T' = 0 \tag{10-10}$$

$$\frac{\partial\sigma}{\partial p}u^2 p' + \left(\frac{\partial\sigma}{\partial T}u^2 - \frac{\sigma^2\rho_s}{\rho_n}\right)T' = 0 \tag{10-11}$$

As usual, the above two equations will be compatible if the determinant of their coefficients is equal to zero. Expanding the determinant we find

$$u^4 \frac{\partial(\sigma,\rho)}{\partial(T,p)} - u^2\left(\frac{\partial\sigma}{\partial T} + \sigma^2 \frac{\rho_s}{\rho_n}\frac{\partial\rho}{\partial p}\right) + \frac{\rho_s}{\rho_n}\sigma^2 = 0 \qquad (10\text{-}12)$$

which after some simple transformations yields

$$u^4 - u^2\left[\left(\frac{\partial p}{\partial\rho}\right)_\sigma + \frac{\rho_s}{\rho_n}\sigma^2\left(\frac{\partial T}{\partial\sigma}\right)_\rho\right] + \frac{\rho_s}{\rho_n}\sigma^2\left(\frac{\partial T}{\partial\sigma}\right)_\rho\left(\frac{\partial p}{\partial\rho}\right)_T = 0 \qquad (10\text{-}13)$$

Equation 10-13 determines two possible velocities of sound in helium II. The coefficient of thermal expansion $(\partial\rho/\partial T)_p$ is, in practice, very small for all substances; for helium II it is even anomalously small. Therefore, according to the well-known thermodynamic relations one can consider the two specific heats c_p and c_v to be practically equal in helium II. But in that case the derivatives $(\partial p/\partial\rho)_T$ and $(\partial p/\partial\rho)_\sigma$ which are related by the equation $(\partial p/\partial\rho)_\sigma = c_p/c_v\,(\partial p/\partial\rho)_T$ can also be considered equal to a high degree of accuracy. This considerably simplifies Eq. 10-13, whose roots become equal to

$$u_1 = c = \sqrt{\left(\frac{\partial p}{\partial\rho}\right)_\sigma} \qquad (10\text{-}14)$$

$$u_2 = \sqrt{\frac{\sigma^2\rho_s}{\rho_n(\partial\sigma/\partial T)}} \qquad (10\text{-}15)$$

The first root determines the velocity of ordinary (first) sound in helium II. From Eq. 10-5 we see that it is with this velocity that oscillations of pressure (density) are propagated in helium II. The second root u_2 determines the velocity of so-called second sound. From Eq. 10-7 we see that oscillations of temperature (entropy) are propagated with this velocity. The ability to propagate undamped temperature waves is a property specific to helium II. The temperature dependence of the velocity of second sound, calculated from formula 10-15, is represented on Fig. 4. At the λ-point $\rho_s = 0$ and the velocity u_2 also goes to zero. At sufficiently low temperatures, when all thermodynamic quantities are determined by the phonons (below 0.5°K), the velocity u_2 tends toward the limit $c/\sqrt{3}$.

Second sound can be looked upon as a compressional wave in the gas of excitations. This follows immediately from the fact that oscillations of temperature bring about oscillations in the density of

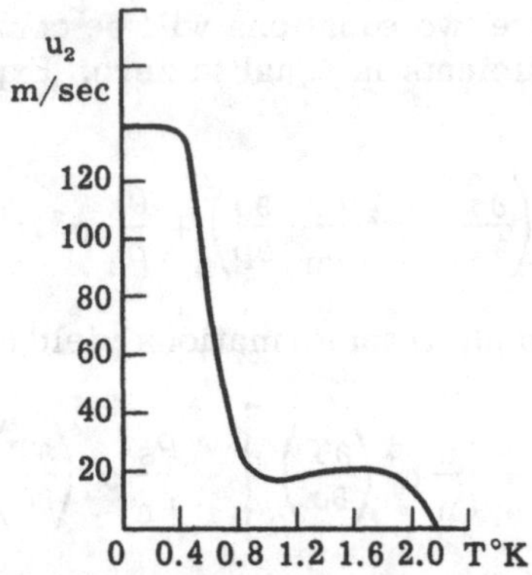

Figure 4. The temperature dependence of the velocity of second sound.

excitations. The velocity of second sound is, therefore, the velocity of ordinary sound in the excitation gas. The limiting value $\mu_2 = c/\sqrt{3}$ follows immediately from the well-known result on the velocity of sound in a gas whose spectrum is given by $\epsilon = cp$.†

Let us examine in more detail the physical nature of these two different sound waves. We consider a plane wave in which all the variables are proportional to

$$\exp\left[i\omega\left(t - \frac{x}{u}\right)\right]$$

If we denote by a prime the varying part of each quantity we obtain from Eqs. 10-1 to 10-4

$$\begin{aligned} &u\rho' - j' = 0 \qquad u\sigma\rho' + u\rho\sigma' - \sigma\rho v_n = 0 \\ &uj' - p' = 0 \qquad uv_s + \sigma T' - \frac{1}{\rho}p' = 0 \end{aligned} \tag{10-16}$$

Let us rewrite this system in a form more convenient for analysis

$$p' = u^2\rho' = uj' \tag{10-17}$$

$$T' = \frac{u^2\rho_n}{\sigma^2\rho_s}\sigma' = \frac{u\rho_n}{\sigma\rho}(v_n - v_s) \tag{10-18}$$

These equations show the relation among the different oscillations

†Thus, for instance, in a gas of photons the velocity of sound is $c/\sqrt{3}$, where c is the velocity of light.

that occur in the two types of sound waves. For each type of sound wave one must insert the corresponding value for the sound velocity u_1 or u_2, from Eqs. 10-14 and 10-15. First sound waves are oscillations of the density ρ and if we neglect the thermal expansion they are not accompanied by any temperature oscillations. According to Eq. 10-18 in this case $\mathbf{v}_n = \mathbf{v}_s$. In other words, in waves of first sound the helium moves as a whole—that is, the normal and superfluid parts oscillate together. First sound waves are clearly analogous to ordinary sound which propagates in normal media.

In waves of second sound it is the temperature and the entropy which oscillate. The pressure and density do not vary in the approximation under consideration. It then follows from Eq. 10-17 that $j' = \rho_s v_s' + \rho_n v_n' = 0$. In this mode the normal and superfluid parts move with respect to one another in such a way that the total current of matter vanishes at each instant. The possibility of sustaining an undamped temperature wave is a specific property of a superfluid, which is due to the existence of two modes of motion. In normal substances temperature waves are damped by thermal conduction in distances of the order of a wavelength ($\sim \sqrt{\chi/\omega}$, where χ is the "temperature conductivity"† and therefore there is no periodicity in such waves.

When thermal expansion is taken into account, first and second sound become coupled, that is there occur temperature oscillations in first sound waves and pressure oscillations in second sound waves. The formulas relating the corresponding amplitudes of oscillations can be obtained from Eqs. 10-17 and 10-18.[14] Let us introduce coefficients of proportionality between the velocities and the variable parts of the temperature and pressure.

$$v_n = a v_s \qquad p' = b v_s \qquad T' = c v_s \tag{10-19}$$

To first order in the coefficient of thermal expansion α we obtain for first sound

$$a_1 = 1 + \frac{\alpha\rho}{\rho_s \sigma}\,\frac{u_1^2 u_2^2}{u_1^2 - u_2^2} \qquad b_1 = \rho u_1 \qquad c_1 = \frac{\rho\alpha T}{C}\,\frac{u_1^3}{u_1^2 - u_2^2} \tag{10-20}$$

and for second sound

$$a_2 = -\frac{\rho_s}{\rho_n} + \frac{\alpha\rho}{\sigma\rho_n}\,\frac{u_1^2 u_2^2}{u_1^2 - u_2^2} \qquad b_2 = \frac{\alpha\rho u_1^2 u_2^2}{\sigma(u_1^2 - u_2^2)} \qquad c_2 = -\frac{u_2}{\sigma} \tag{10-21}$$

†The "temperature conductivity" χ is defined as $\chi \sim (K/C)^2 (\omega/c_1^2)$, where K is the thermal conductivity, C the heat capacity, and c_1 the velocity of sound.

11

THE EXCITATION OF SOUND WAVES IN A SUPERFLUID[14,15]

Since first and second sound have different physical natures it is natural that they will be excited by different means. Let us consider some examples of the excitation of sound waves. We first look at the excitation of sound waves by a plane oscillating in a direction perpendicular to itself (which we choose as the x-axis). We may write the x-component of the velocity v_s in the first and second type of excited sound waves in the form

$$v_{s1} = A_1 \exp\{-i\omega[t - (x/u)]\}$$
$$v_{s2} = A_2 \exp\{-i\omega[t - (x/u)]\} \tag{11-1}$$

From the boundary condition Eq. 8-36 it follows that at the boundary of the solid body the velocities v_{sx} and v_{nx} coincide with the velocity $v_0 e^{-i\omega t}$ of the surface. We thus have

$$A_1 + A_2 = 0 \qquad A_1 a_1 + A_2 a_2 = 0$$

where a_1, a_2 are the coefficients in formula 10-19. We then obtain

$$\frac{A_2}{A_1} = -\frac{1 - a_1}{1 - a_2} = \frac{\alpha\rho_n u_2^2}{\rho_s \sigma} = \frac{T\sigma\alpha p}{C} \tag{11-2}$$

We here used Eqs. 10-20 and 10-21 which are simplified at not too low temperatures where we may neglect u_2^2 in comparison with u_1^2.

Since during oscillatory motion the average kinetic energy is equal to the average potential energy, the total energy density is equal to

$$\overline{E} = \rho_s \overline{v_s^2} + \rho_n \overline{v_n^2} = \tfrac{1}{2}|A|^2(\rho_s + \rho_n a^2) \tag{11-3}$$

The energy flux (intensity) is obtained by multiplying this density by the corresponding velocity of sound. The ratio of the intensities of the waves of first and second sound thus excited is

$$\frac{I_2}{I_1} = \frac{u_2|A_2|^2(\rho_s + \rho_n a_2^2)}{u_1|A_1|^2(\rho_s + \rho_n a_1^2)}$$

$$\approx \frac{\alpha^2 T u_2^3 \rho}{C u_1} \qquad (a_1 \approx 1,\ a_2 \approx -\rho_s/\rho_n) \tag{11-4}$$

Let us likewise write the expressions for I_1 and I_2 in terms of the oscillation amplitudes of the pressure p_0' and the temperature T_0'

$$I_1 = \frac{1}{2\rho u_1} p_0'^2 \qquad I_2 = \frac{C u_2}{2T} T_0'^2 \tag{11-5}$$

The intensity of the second sound wave thus excited turns out to be infinitesimally small, since it is proportional to the square of the thermal expansion. A plane oscillating in this manner excites primarily first sound. This is easy to understand if we remember that at the surface of a solid body $v_s = v_n$, whereas second sound is connected with the possibility of a nonzero difference $v_n - v_s$.

One may create relative motion of the superfluid and normal parts by immersing into the helium a body whose temperature varies periodically. Let us consider, for instance, the excitation of sound by a fixed solid plane whose temperature varies periodically.

The boundary conditions at the surface are in this case that the normal component of the matter current j should be zero and that the difference in temperature between the solid and the liquid should be proportional to the density of heat flux. In order to calculate the intensity of the sound thus obtained it is sufficient to take into account the first of these boundary conditions, which states

$$\rho_s(A_1 + A_2) + \rho_n(a_1 A_1 + a_2 A_2) = 0$$

whence

$$\left|\frac{A_2}{A_1}\right| - \frac{\rho_n a_1 + \rho_s}{\rho_n a_2 + \rho_s} \sim \frac{\sigma}{\alpha u_2^2}$$

The ratio of the intensities is

$$\frac{I_2}{I_1} = \frac{C}{\rho T \alpha^2 u_1 u_2} \tag{11-6}$$

This ratio is very large (greater than 10^3), so that in practice we only excite second sound in this case.

When sound in liquid helium is reflected by the boundary between the liquid and its vapor, we have the curious phenomenon of transformation of second sound into first sound and vice versa. Indeed, when a second sound wave propagating in helium II is reflected at the liquid-gas interface, it creates temperature oscillations which cause periodic condensation and vaporization of the gas. As a result, there occur periodic variations of density in the vapor near the surface, which propagate into the vapor in the form of ordinary sound waves.

Let us now study the case of a second sound wave T_i' incident on the surface between helium II and its vapor (see Fig. 5). There appear two reflected waves in the liquid, T_r' (a reflected wave of second sound) and p' (a first sound wave), and one wave in the vapor, $\overline{p}'$, an adiabatic wave. We shall suppose that we may neglect the thermal conductivity of the vapor. This is always possible for sound frequencies satisfying the condition $\omega \ll \overline{c}^2/\overline{\chi}$ ($\overline{c}$ is the adiabatic sound velocity, $\overline{\chi}$ the "temperature conductivity" of the vapor). Nevertheless in the boundary conditions, as we shall see, it is important to take into

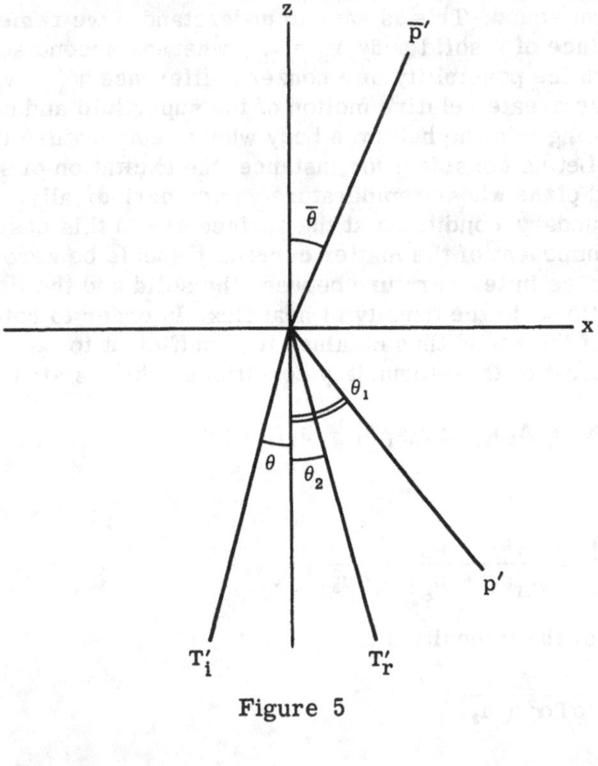

Figure 5

account the temperature wave $\overline{T}'$, even though this wave is rapidly damped.

Let us take the plane xy as our boundary plane. From the homogeneity of the problem in this plane we see that all waves will have the same wave-vectors K_x and K_y. From this there immediately follows a relation determining the direction of propagation of the waves which are excited. Let xz be the plane of incidence of the wave T_i' and θ, θ_1, θ_2, $\overline{\theta}$ the angles between the z-axis and the directions of propagation of the waves T_i', p', T_r', $\overline{p}'$, respectively. Then from the equality of K_x and K_y it follows that all waves lie in the same plane

$$\sin\theta_1 = \frac{u_1}{u_2}\sin\theta \qquad \theta_2 = \theta \qquad \sin\overline{\theta} = \frac{\overline{c}}{u_2} \tag{11-7}$$

u_1 and u_2 are the velocities of first and second sound, respectively, and $\overline{c}$ the velocity of sound in the vapor.

Let us introduce the quantity

$$\zeta(x,t) = \zeta(o)\exp[-i\omega t + ix(\omega/u_2)\sin\theta]$$

characterizing the oscillations of the interface. We now write the conditions which must be satisfied on this interface. We denote by the index o the amplitude of a quantity, by a bar its value in the vapor and by the symbol without a bar its value in the liquid.

(a) The force equation on the boundary

$$p^{(o)} - \overline{p}^{(o)} = \gamma\frac{\omega^2}{u_2^2}\sin^2\theta\,\zeta^{(o)} \tag{11-8}$$

γ is the coefficient of surface tension.

(b) The equality of the matter current densities on both sides of the boundary

$$-\frac{\cos\theta_1}{u_1}p^{(o)} + i\omega\rho\zeta^{(o)} = \frac{\cos\overline{\theta}}{\overline{c}}p^{(o)} + i\omega\overline{\rho}\zeta^{(o)} \tag{11-9}$$

(c) The equality of the energy-flux densities

$$W\left(-\frac{\cos\theta_1}{u_1}p^{(o)} + i\omega\rho\zeta^{(o)}\right) + \rho\frac{\rho_s}{\rho_n}\frac{\sigma^2 T}{u_2}\cos\theta\left(T_i^{(o)} - T_r^{(o)}\right)$$

$$= \overline{W}\left(\frac{\cos\overline{\theta}}{\overline{c}}\overline{p}^{(o)} + i\omega\rho\zeta^{(o)}\right) \tag{11-10}$$

where W is the heat function and σ the entropy.

(d) The equality of temperatures

$$T_i^{(o)} + T_r^{(o)} = \frac{T\alpha}{\overline{C}_p}\,\overline{p}^{(o)} + \overline{T}^{(o)} \tag{11-11}$$

α is the coefficient of thermal expansion of the vapor.

(e) The equality of chemical potentials

$$\frac{1}{\rho}p^{(o)} - \sigma\left(T_i^{(o)} + T_r^{(o)}\right) = \frac{1}{\overline{\rho}}\overline{p}^{(o)} - \overline{\sigma}\left(\frac{\alpha T}{\overline{C}_p}\overline{p}^{(o)} + \overline{T}^{(o)}\right) \tag{11-12}$$

In the frequency interval which interests us we may neglect the surface tension ($\gamma\omega/\rho u_1 u_2^2 \ll 1$) in Eq. 11-8. Furthermore if we note that $\sigma/\overline{\sigma} \ll 1$, $\overline{\rho}/\rho \ll 1$, the system, Eqs. 11-8 and 11-12, simplifies and we find

$$\frac{\overline{p}^{(o)}}{T_i^{(o)}} = \frac{p^{(o)}}{T_i^{(o)}} = \frac{2\rho\sigma\dfrac{\rho_s}{\rho_n}\dfrac{\sigma}{\overline{\sigma}}\dfrac{\overline{c}}{u_2}\cos\theta}{\cos\overline{\theta} + \dfrac{\rho_s}{\rho_n}\dfrac{\rho}{\overline{\rho}}\left(\dfrac{\sigma}{\overline{\sigma}}\right)^2\dfrac{\overline{c}}{u_2}\cos\theta} \tag{11-13}$$

$$\frac{T_r^{(o)}}{T_i^{(o)}} = \frac{-\cos\overline{\theta} + \dfrac{\rho_s}{\rho_n}\dfrac{\rho}{\overline{\rho}}\left(\dfrac{\sigma}{\overline{\sigma}}\right)^2\dfrac{\overline{c}}{u_2}\cos\theta}{\cos\overline{\theta} + \dfrac{\rho_s}{\rho_n}\dfrac{\rho}{\overline{\rho}}\left(\dfrac{\sigma}{\overline{\sigma}}\right)^2\dfrac{\overline{c}}{u_2}\cos\theta} \tag{11-14}$$

where

$$\cos\overline{\theta} = \sqrt{1 - \frac{\overline{c}^2}{u_2^2}\sin^2\theta}$$

By using formulas 11-5, 11-13, and 11-14, we may calculate the reflection coefficient of the second sound R_2, the coefficient of transformation of second sound into first sound R_1, and into ordinary sound in the vapor $\overline{R}$.

$$R_2 = \left(\frac{-\cos\overline{\theta} + \dfrac{\rho_s}{\rho_n}\dfrac{\rho}{\overline{\rho}}\left(\dfrac{\sigma}{\overline{\sigma}}\right)^2\dfrac{\overline{c}}{u_2}\cos\theta}{\cos\overline{\theta} + \dfrac{\rho_s}{\rho_n}\dfrac{\rho}{\overline{\rho}}\left(\dfrac{\sigma}{\overline{\sigma}}\right)^2\dfrac{\overline{c}}{u_2}\cos\theta}\right)^2 \tag{11-15}$$

$$R_1 = \frac{4\dfrac{\rho_s}{\rho_n}\left(\dfrac{\sigma}{\overline{\sigma}}\right)^2 \dfrac{\overline{c}}{u_2}\dfrac{\overline{c}}{u_1} \cos\theta \cos\theta_1}{\left(\cos\overline{\theta} + \dfrac{\rho_s}{\rho_n}\dfrac{\rho}{\overline{\rho}}\left(\dfrac{\sigma}{\overline{\sigma}}\right)^2 \dfrac{\overline{c}}{u_2}\cos\theta\right)^2} \tag{11-16}$$

$$\overline{R} = \frac{4\dfrac{\rho_s}{\rho_n}\dfrac{\rho}{\overline{\rho}}\left(\dfrac{\sigma}{\overline{\sigma}}\right)^2 \dfrac{\overline{c}}{u_2} \cos\theta \cos\overline{\theta}}{\left(\cos\overline{\theta} + \dfrac{\rho_s}{\rho_n}\dfrac{\rho}{\overline{\rho}}\left(\dfrac{\sigma}{\overline{\sigma}}\right)^2 \dfrac{\overline{c}}{u_2}\cos\theta\right)^2} \tag{11-17}$$

From Eq. 11-17 it may be seen that for incident angles $\theta >$ arcsin u_2/u_1 there is no reflected first sound wave in the liquid and for angles $\theta >$ arcsin $u_2/\overline{c}$ the sound wave in the vapor disappears. In the last case we have complete internal reflection of the second sound wave. This complete reflection occurs for the temperatures greater than 0.7°K, when $\overline{c} > u_2$.

It should be noted that the coefficient R_1, determining the transformation of second sound into first sound, is always small in helium II.

12

THE ABSORPTION OF SOUND[16]

The presence of dissipative processes in a superfluid leads to absorption or damping of sound. In order to investigate the question of sound propagation in the presence of dissipation let us write the general equations, 9-13 to 9-16, in the following linearized form:

$$\dot{\rho} + \operatorname{div} \mathbf{j} = 0 \tag{12-1}$$

$$\dot{j}_i + \frac{\partial p}{\partial r_i} = \eta \frac{\partial}{\partial r_k}\left(\frac{\partial v_{ni}}{\partial r_k} + \frac{\partial v_{nk}}{\partial r_i} - \frac{2}{3}\,\delta_{ik}\frac{\partial v_{n\ell}}{\partial r_\ell}\right)$$

$$+ \frac{\partial}{\partial r_i}\{\zeta_1 \operatorname{div}(\mathbf{j} - \rho \mathbf{v}_n) + \zeta_2 \operatorname{div} \mathbf{v}_n\} \tag{12-2}$$

$$\dot{\mathbf{v}}_s + \nabla\mu = \nabla\{\zeta_3 \operatorname{div}(\mathbf{j} - \rho\mathbf{v}_n) + \zeta_4 \operatorname{div} \mathbf{v}_n\} \tag{12-3}$$

$$T\{\dot{(\sigma\rho)} + \sigma\rho \operatorname{div} \mathbf{v}_n\} = \kappa \Delta T \tag{12-4}$$

In a sound wave the velocities v_n and v_s and the varying parts of the thermodynamic variables ρ' and σ' (which we choose as our independent variables) vary as $\exp[-i\omega(t - x/u)]$ (x is the direction of propagation and ω the frequency of the sound wave). The velocity of sound will in this case be a complex quantity u, its imaginary part determining the damping. In order to simplify the calculations let us use the fact that the thermal expansion in helium II is very small. By eliminating the variables v_n and v_s from Eqs. 12-1 to 12-4, we obtain

$$\left(u^2 - \frac{\partial p}{\partial \rho}\right)\rho' = i\omega\left\{\left(\tfrac{4}{3}\eta + \zeta_2\right)\frac{\rho'}{\rho} + \left(\tfrac{4}{3}\eta + \zeta_2 - \rho\zeta_1\right)\frac{\sigma'}{\sigma}\right\} \tag{12-5}$$

$$\left(\sigma \frac{\partial T}{\partial \sigma} - \frac{\rho_n}{\rho_s \sigma} u^2\right)\sigma'$$

$$= i\omega \left\{\left(\zeta_4 - \frac{1}{\rho}\zeta_2 - \frac{4}{3\rho}\eta\right)\frac{\rho'}{\rho} + \left(\zeta_4 - \rho\zeta_3 - \frac{1}{\rho}\zeta_2 + \zeta_1 - \frac{4}{3\rho}\eta\right) \times \frac{\sigma'}{\sigma} - \frac{\rho_n}{\rho_s}\frac{\kappa}{\rho\sigma T}\frac{\partial T}{\partial \sigma}\sigma'\right\} \tag{12-6}$$

These two equations will be compatible if the determinant of their coefficients vanishes. For low frequencies we may limit ourselves to terms linear in ω. But as we may easily convince ourselves, this means that we can neglect the terms containing σ' in Eq. 12-5 and the terms containing ρ' in Eq. 12-6. As a result we obtain two independent homogeneous equations from which the equation for the velocity of sound follows

$$u^2 - \left(\frac{\partial p}{\partial \rho}\right)_\sigma = \frac{i\omega}{\rho}\left(\tfrac{4}{3}\eta + \zeta_2\right) \tag{12-7}$$

$$\left(\sigma \frac{\partial T}{\partial \sigma} - \frac{\rho_n}{\rho_s \sigma} u^2\right) = \frac{i\omega}{\rho\sigma}\left\{\rho(\zeta_1 + \zeta_4) - \rho^2\zeta_3 - \zeta_2 - \tfrac{4}{3}\eta - \frac{\rho_n}{\rho_s}\frac{\kappa}{T}\frac{\partial T}{\partial \sigma}\right\} \tag{12-8}$$

The root of Eq. 12-7 determines the velocity of first sound, when damping is taken into account

$$u_1^2 = \left(\frac{\partial p}{\partial \rho}\right)_\sigma + \frac{i\omega}{\rho}\left(\tfrac{4}{3}\eta + \zeta_2\right) \tag{12-9}$$

The root of Eq. 12-8 determines the velocity of second sound, also including damping

$$u_2^2 = \sigma^2 \frac{\partial T}{\partial \sigma}\frac{\rho_s}{\rho_n} + \frac{i\omega}{\rho}\frac{\rho_s}{\rho_n}\left\{\zeta_2 + \rho^2\zeta_3 - 2\rho\zeta_1 + \tfrac{4}{3}\eta + \frac{\rho_n}{\rho_s}\frac{\kappa}{T}\frac{\partial T}{\partial \sigma}\right\} \tag{12-10}$$

The velocities u_1 and u_2 are complex quantities and therefore the wave vectors will also be complex. The real part of the wave vector determines the change in the phase with distance and the imaginary part is the coefficient of sound absorption. For first sound it is equal to

$$\alpha_1 = \operatorname{Im} \frac{\omega}{u_1} = \frac{\omega^2}{2\rho u_1^3}\left(\tfrac{4}{3}\eta + \zeta_2\right) \tag{12-11}$$

Thus the coefficient of absorption of first sound depends only on two kinetic coefficients, the first viscosity η and the second viscosity ζ_2. The other coefficients do not appear in our expression because we neglected the thermal expansion of helium II. Similarly the thermal conductivity leads to an additional term in α_1 of the form

$$\frac{\omega^2}{2u_1^3}\,\frac{\kappa}{C}\left(\frac{C_p}{C_V} - 1\right) \tag{12-12}$$

We may note that the most important contribution to the absorption of first sound comes from the coefficient ζ_2, as can be seen by calculating η and ζ_2.

The coefficient of absorption of second sound is equal to the imaginary part of the wave vector calculated with the aid of expression 12-10

$$\begin{aligned}\alpha_2 &= \operatorname{Im} \frac{\omega}{u_2} \\ &= \frac{\omega^2}{2\rho u_2^3}\,\frac{\rho_s}{\rho_n}\left\{\tfrac{4}{3}\eta + (\zeta_2 + \rho^2\zeta_3 - 2\rho\zeta_1) + \frac{\rho_n}{\rho_s}\,\frac{\kappa}{T}\,\frac{\partial T}{\partial \sigma}\right\}\end{aligned} \tag{12-13}$$

A detailed analysis shows that the major contribution comes from the thermal conductivity, whose effect is described by the last term in Eq. 12-13.

13

DISCONTINUITIES IN A SUPERFLUID[17]

As is known from hydrodynamics, when sound waves of large amplitude are propagated in a liquid, there appear shock waves with surfaces of discontinuity. The essence of this phenomenon is the following: different points of the wave front move with different velocities, and this leads to a deformation of the wave front in time. Points with large velocities move forward (to the crest of the wave), leaving behind them the points with smaller velocities. Finally the wave front may be bent so much that it ceases to be a single-valued function of the coordinates. Physically, such a situation is not possible. What in fact happens is that a discontinuity appears, which separates off one part of the twisted wave front. As a result all physical quantities in the wave turn out to be single-valued functions of the coordinates. The shock waves which are thus formed from sound waves, naturally have small amplitudes.

A similar phenomenon should occur for first and second sound waves in helium II. Discontinuities in first sound waves should, of course, correspond completely to the usual shock waves. In these, the quantities having discontinuities should be primarily the pressure, the density and the velocities $\mathbf{v}_n = \mathbf{v}_s$. The discontinuity in the entropy and the associated one in the temperature and relative velocity (neglecting the thermal expansion) will be small quantities of third order.

Discontinuities in second sound waves are specific to superfluids; they are primarily discontinuities in the temperature, along with the relative velocity $\mathbf{w}$. The discontinuities in the remaining thermodynamic quantities are of higher order. Let us study the conditions which should be satisfied at the surfaces of discontinuity in helium II.

To do this, let us introduce a frame of reference moving with the discontinuity (at a velocity u in the rest frame). We choose the x-axis to be along the normal to the surface of discontinuity, and the velocities of motion of the liquid along the x-axis.†

The following quantities should be constant on the surface of discontinuity:

(a) The current density of the liquid

$$[j_x] = [\rho_s v_{sx} + \rho_n v_{nx}] = 0 \tag{13-1}$$

(in the last equation and henceforth the square brackets will denote the difference in the values of the corresponding quantity on both sides of the surface).

(b) The momentum current density

$$[\Pi_{xx}] = [\rho_s v_{sx}^2 + \rho_n v_{nx}^2 + p] = 0 \tag{13-2}$$

(c) The quantity under the divergence sign in the equation of superfluid motion, Eq. 8-25, namely the force acting on unit mass of the superfluid part of the system,

$$\left[\tilde{\mu} + \frac{v_s^2}{2}\right] = 0 \tag{13-3}$$

(the notation ~ over a quantity means that we consider it as a function of p, T, and w).

(d) The energy density current; according to Eq. 8-20 this condition is

$$\left[j_x\left(\tilde{\mu} + \frac{v_s^2}{2}\right) + T\tilde{\rho}\tilde{\sigma} v_{nx} + \rho_n (\mathbf{v}_n \cdot \mathbf{w})\, v_{nx}\right] = 0$$

but since the first term is continuous by Eqs. 13-2 and 13-3 it is enough to require that

$$[T\tilde{\rho}\tilde{\sigma} v_{nx} + \rho_n (\mathbf{v}_n \cdot \mathbf{w})\, v_{nx}] = 0 \tag{13-4}$$

Let us now return to the rest frame. In order to simplify the ensuing calculations we shall assume that on one side of the discontinuity we have the unperturbed liquid, in other words that $v_n = v_s = 0$ and all quantities are equal to their equilibrium values. We shall consider the velocity u of the discontinuity to be positive if the latter moves toward the unperturbed liquid. Then in Eqs. 13-1 to 13-4 we

†We shall not consider here tangential discontinuities, for which the components of velocity tangential to the surface may have jumps.

must set the velocities v_s and v_n in the frame of reference moving with the discontinuity equal to u on the unperturbed side. On the other side, they must be changed to $v_s - u$ and $v_n - u$, and now v_s and v_n will refer to the rest frame. It is convenient to introduce, along with the velocity difference $w = v_n - v_s$, the velocity v which is formally related to the current density by the equation $j = \tilde{\rho} v$. As a result the set of conditions Eqs. 13-1 to 13-4 takes the form

$$\rho_0 u = \tilde{\rho}(u - v) \tag{13-5}$$

$$p_0 + \rho_0 u^2 = p + \tilde{\rho}(u - v)^2 + \frac{\rho_s \rho_n}{\tilde{\rho}} w^2 \tag{13-6}$$

$$\mu_0 + \frac{u^2}{2} = \tilde{\mu} + \frac{1}{2}\left(u - v + \frac{\rho_n}{\tilde{\rho}} w\right)^2 \tag{13-7}$$

$$\rho_0 T_0 \sigma_0 u = \tilde{\rho}\tilde{\sigma} T\left(u - v - \frac{\rho_s}{\tilde{\rho}} w\right) + \rho_n w \left(u - v - \frac{\rho_s}{\tilde{\rho}} w\right)^2 \tag{13-8}$$

(the index 0 on a quantity denotes the equilibrium value of the quantity; the symbol with no index denotes the value of the quantity on the other side of the discontinuity).

The above system may in principle be used to find the velocity of shock waves and the magnitude of the discontinuities in the physical quantities at the surface. However, because of the complicated dependence of the thermodynamic variables on the relative velocity w, a general analysis would be extremely difficult. We shall limit ourselves to a consideration of discontinuities of small intensity, for which it is sufficient to expand the thermodynamic variables $\tilde{\mu}$, $\tilde{\sigma}$, and $\tilde{\rho}$ to second order in w.

Let us use Eq. 13-5 to eliminate $u - v$ from the other equations and insert into these expressions 8-26, 8-28, and 8-29 for $\tilde{\mu}$, $\tilde{\sigma}$, and $\tilde{\rho}$. This yields

$$p - p_0 - \frac{\rho_0}{\rho} u^2 (\rho - \rho_0) + w^2 \rho_0 \left[\frac{\rho_s \rho_n}{\rho^2} - \frac{\rho_0 u^2}{2} \frac{\partial}{\partial p} \frac{\rho_n}{\rho}\right] = 0 \tag{13-9}$$

$$\mu - \mu_0 - \frac{u^2}{2\rho^2}(\rho^2 - \rho_0^2) + \frac{\rho_0 \rho_n}{\rho^2} wu - w^2 \left[\frac{\rho_s \rho_n}{2\rho^2} + \frac{\rho_0 u^2}{2} \frac{\partial}{\partial p} \frac{\rho_n}{\rho}\right] = 0 \tag{13-10}$$

$$\rho_0 u (T\sigma - T_0 \sigma_0) - w\left(\sigma T \rho_s + \frac{\rho_n \rho_0^2}{\rho^2} u^2\right) + \rho_0 w^2 u \left[\frac{2\rho_n \rho_s}{\rho^2} + \tfrac{1}{2} T \frac{\partial}{\partial T} \frac{\rho_n}{\rho}\right] = 0 \tag{13-11}$$

For what follows let us choose as our independent variables the pressure p and the temperature T and expand all quantities in powers of $\Delta p = p - p_0$ and $\Delta T = T - T_0$, limiting ourselves to quadratic terms. Furthermore, we neglect the dependence of the density ρ on the temperature, as we have already done up to now. The system, Eqs. 13-9 to 13-11, then becomes

$$\Delta p\left(1 - u^2 \frac{\partial \rho}{\partial p}\right) + (\Delta p)^2 u^2 \left[\frac{1}{\rho}\left(\frac{\partial \rho}{\partial p}\right)^2 - \frac{1}{2}\frac{\partial^2 \rho}{\partial p^2}\right]$$

$$+ w^2\left[\frac{\rho_s \rho_n}{\rho} - \tfrac{1}{2}\rho^2 u^2 \frac{\partial}{\partial p}\frac{\rho_n}{\rho}\right] = 0 \qquad (13\text{-}12)$$

$$\frac{1}{\rho}\,\Delta p\left(1 - u^2\frac{\partial \rho}{\partial p}\right) + (\Delta p)^2\left[-\frac{1}{2\rho^2}\frac{\partial \rho}{\partial p} + \frac{3}{2\rho^2}\left(\frac{\partial \rho}{\partial p}\right)^2 u^2 - \frac{1}{2\rho}u^2\frac{\partial^2 \rho}{\partial p^2}\right]$$

$$-\sigma\Delta T + \frac{\rho_n}{\rho}uw + \Delta p w \frac{1}{\rho}\left[\frac{\partial \rho_n}{\partial p} - 2\frac{\rho_n}{\rho}\cdot\frac{\partial \rho}{\partial p}\right]$$

$$+ w^2\left[-\frac{\rho_n \rho_s}{2\rho^2} - \tfrac{1}{2}\rho u^2\frac{\partial}{\partial p}\frac{\rho_n}{\rho}\right] - \tfrac{1}{2}(\Delta T)^2\frac{\partial \sigma}{\partial T}$$

$$+ \Delta T\, wu \frac{\partial}{\partial T}\frac{\rho_n}{\rho} = 0 \qquad (13\text{-}13)$$

$$\Delta T\,\rho u\left(\sigma + T\frac{\partial \sigma}{\partial T}\right) - w(\sigma T \rho_s + \rho_n u^2) + (\Delta T)^2 \rho u\left(\frac{\partial \sigma}{\partial T} + \frac{1}{2}T\frac{\partial^2 \sigma}{\partial T^2}\right)$$

$$- w\,\Delta T\left[u^2\frac{\partial \rho_n}{\partial T} + \sigma\rho_s + T\frac{\partial}{\partial T}(\sigma\rho_s)\right]$$

$$- w\,\Delta p\left(-\frac{\rho_n}{\rho}\frac{\partial \rho}{\partial p}u^2 + \frac{\partial \rho_n}{\partial p}u^2 + T\sigma\frac{\partial \rho_s}{\partial p}\right)$$

$$+ w^2 u\left(T\frac{\partial \rho_n}{\partial T} + 2\rho_s\rho_n\right) = 0 \qquad (13\text{-}14)$$

In linear approximation this system simplifies considerably. The compatibility condition is

$$\left(1 - u^2\frac{\partial \rho}{\partial p}\right)\left(\frac{\partial \sigma}{\partial T}u^2\rho_n - \sigma^2\rho_s\right) = 0 \qquad (13\text{-}15)$$

which expresses the value of the velocities u to first approximation. The roots of Eq. 13-15 are

$$u_{10}^2 = \frac{\partial p}{\partial \rho} \tag{13-16}$$

$$u_{20}^2 = \frac{\rho_s}{\rho_n} \sigma^2 \frac{\partial T}{\partial \sigma} \tag{13-17}$$

The result thus obtained is rather obvious: weak discontinuities in the pressure or temperature propagate at the velocities of the corresponding sounds.

If we insert the values of the velocities u thus obtained into the system, Eqs. 13-12 to 13-14, we find the relation between the discontinuities Δp, ΔT, and w at the surface. In this manner we may convince ourselves that the discontinuities ΔT and w which correspond to the first root, are of higher than first order compared to the discontinuity Δp. Similarly the discontinuity Δp corresponding to the second root is of higher than first order compared to ΔT and w. Furthermore in the latter case the following relation between ΔT and w may be found

$$\Delta T = w u_2 \frac{\rho_n}{\rho \sigma} \tag{13-18}$$

The first root corresponds to pressure discontinuities which are analogous to shock waves in normal media. The second root corresponds to a temperature discontinuity.

PRESSURE DISCONTINUITIES (SHOCK WAVES)

Let us neglect terms which are higher than second order in Eq. 13-11. We get

$$\left(1 - u^2 \frac{\partial \rho}{\partial p}\right) + \left(\frac{1}{\rho}\left(\frac{\partial \rho}{\partial p}\right)^2 - \frac{1}{2}\frac{\partial^2 \rho}{\partial p^2}\right) u^2 \, \Delta p = 0 \tag{13-19}$$

which determines the velocity of the discontinuity to second order

$$u_1 = u_{10}\left\{1 + \Delta p \frac{\partial}{\partial p} \ln (\rho u_{10})\right\} \tag{13-20}$$

According to Eq. 13-1, for small values of the jump in velocity at the surface we have

$$\Delta p = \rho u_{10} v$$

By using this relation, we may rewrite Eq. 13-20 in the form

$$u_1 = u_{10}\left\{1 + \tfrac{1}{2}v\,\frac{\partial}{\partial p}(\rho u_{10})\right\} \tag{13-21}$$

Equation 13-21 coincides with the expression for the velocity of the discontinuity for normal media. Inserting the value of u^2 thus obtained into the two remaining equations we may convince ourselves that in this case the discontinuities ΔT and w turn out to be of higher than second order compared to the discontinuity in pressure Δp.

Thus pressure discontinuities in helium II are completely equivalent to shock waves usually encountered in hydrodynamics. The sign of $\partial/\partial p \ln (\rho u_{10})$ is also positive, just as for normal media. Therefore shock waves of this type may only be compression waves. The first term in Eq. 13-20 is equal to the velocity of sound in helium II. Thus in the first approximation pressure discontinuities propagate with the velocity of first sound.

TEMPERATURE DISCONTINUITIES

The second root u_{20} corresponds to discontinuities ΔT and w which are nonzero in first approximation. Let us neglect terms of higher than the second order in Eqs. 13-12 to 13-14. Equation 13-12 determines the pressure jump Δp in terms of the velocity w. This discontinuity turns out to be of second order relative to w (or ΔT).

$$\Delta p = -w^2\left[\frac{\rho_s \rho_n}{\rho} - \tfrac{1}{2}\rho^2 u_{20}^2\,\frac{\partial}{\partial p}\left(\frac{\rho_n}{\rho}\right)\right] \tag{13-22}$$

Let us insert this expression for Δp into Eq. 13-13 and eliminate the velocity w from Eqs. 13-12 and 13-14. This yields

$$\rho\left(-\sigma^2\rho_2 + u^2\,\frac{\partial\sigma}{\partial T}\rho_n\right) + \rho_s\sigma T(\Delta T)^2 \times \left\{-3\,\frac{\partial\sigma}{\partial T} + \tfrac{3}{2}\sigma\,\frac{\rho}{\rho_n\rho_s}\,\frac{\partial\rho_n}{\partial T} - \tfrac{1}{2}\sigma\,\frac{\partial T}{\partial\sigma}\,\frac{\partial^2\sigma}{\partial T^2}\right\} = 0 \tag{13-23}$$

which determines the velocity u_2 in second order. Solving Eq. 13-23 for u and taking into account the definition Eq. 13-17 we have

$$u_2 = u_{20}\left\{1 + \frac{1}{2}\,\frac{\partial}{\partial T}\ln\left(u_{20}^3\,\frac{\partial\sigma}{\partial T}\right)\Delta T\right\} \tag{13-24}$$

This may be rewritten in the form

$$u_2 = u_{20}\left\{1 + \frac{1}{2}\,\frac{\rho_n u_{20}^2}{\rho\sigma}\,w\,\frac{\partial}{\partial T}\ln\left(u_{20}^3\,\frac{\partial\sigma}{\partial T}\right)\right\} \tag{13-25}$$

Temperature discontinuities are a phenomenon specific to superfluids. The velocity of such discontinuities is determined by Eq. 13-25. In first approximation it is equal to the velocity of second sound u_{20}. The sign of the derivative $\partial/\partial T \ln [u_{20}^3 (\partial\sigma/\partial T)]$ changes with changing temperature; this leads to rather curious phenomena.

Different points on the wave front of a second sound wave move with different velocities. This leads to a deformation of the wave front in time. When the wave front ceases to be a single-valued function of the coordinates, discontinuities appear (in this case temperature discontinuities). The velocity of points on the wave front changes discontinuously. The velocity of the discontinuity thus formed depends on the magnitude of the jump in the velocity of the particles. According to Eq. 13-25 it is equal to†

$$u_2 = u_{20} + \tfrac{1}{2}(w_1 + w_2)\,\frac{\rho_s \sigma}{\rho}\,\frac{\partial T}{\partial \sigma}\,\frac{\partial}{\partial T}\ln\left(u_{20}^3 \frac{\partial \sigma}{\partial T}\right) \qquad (13\text{-}26)$$

(w_1 and w_2 are the values of the relative velocity on either side of the surface of discontinuity).

In second sound waves the velocity w is related to the velocity of normal motion by the equation $w = v_n(\rho/\rho_s)$. We may use this to rewrite Eq. 13-26 as

$$u_2 = u_{20} + \tfrac{1}{2}\alpha_2 (v_{n_1} + v_{n_2}) \qquad (13\text{-}27)$$

where

$$\alpha_2 = \frac{ST}{C}\,\frac{\partial}{\partial T}\ln\left(u_{20}^3 \frac{C}{T}\right) \qquad (13\text{-}28)$$

The coefficient α_2 changes sign as the temperature changes. Its dependence on the temperature is depicted in the figure of Ref. 17. For temperatures above 2.0°K, as well as in the interval from 0.4 to 0.9°K the coefficient α_2 is negative. In the remaining temperature range, α_2 is positive. For temperatures such that α_2 is negative the surface of discontinuity occurs at the back of the wave, and for temperatures where α_2 is positive it arises at the front. The appearance of a discontinuity at the back of the wave is a property which is specific to second sound in helium II and is unknown for normal sound.

†In deriving Eq. 13-25 we assumed that the velocity ω was zero on one side of the surface. In the general case, one obtains the formula which is given in the text.

14

FOURTH SOUND[18]

In narrow capillaries it is possible to have a situation in which the wavelength of the excitations becomes comparable to or greater than the diameter of the pipe. In that case, when helium flows, the normal part is stationary. It is then possible to have sound propagation in the superfluid part; these oscillations are called fourth sound. The velocity of fourth sound can be found from the linearized system of hydrodynamic equations in which one sets $\mathbf{v}_n = 0$. According to Eqs. 8-22 to 8-25 we have

$$\dot{\rho} + \rho \operatorname{div} \mathbf{v}_s = 0 \qquad (14\text{-}1)$$

$$\dot{\mathbf{v}}_s + \nabla\mu = 0 \qquad (14\text{-}2)$$

$$\dot{S} = 0 \qquad (14\text{-}3)$$

If we eliminate from the first two equations the velocity $\mathbf{v}_s$ and use the identity, Eq. 8-26, for μ, we obtain

$$\ddot{\rho} = \frac{\rho_s}{\rho}\,\Delta p - \sigma\rho_s\,\Delta T \qquad (14\text{-}4)$$

If we neglect the coefficient of thermal expansion $\partial\rho/\partial T$, Eq. 14-3 yields

$$\rho\,\frac{\partial\sigma}{\partial T}\,\Delta T + \sigma\,\frac{\partial\rho}{\partial p}\,\Delta p = 0 \qquad (14\text{-}5)$$

From Eqs. 14-4 and 14-5 we obtain a wave equation for p

$$\ddot{p} - \left[\frac{\rho_s}{\rho}\frac{\partial p}{\partial \rho} + \frac{\rho_s \sigma}{\rho(\partial\sigma/\partial T)}\right] \Delta p = 0 \tag{14-6}$$

which possesses a periodic solution propagating with the velocity

$$u_4^2 = \frac{\rho_s}{\rho} u_1^2 + \frac{\rho_n}{\rho} u_2^2 \tag{14-7}$$

where

$$u_1 = \sqrt{\frac{\partial p}{\partial \rho}}$$

is the velocity of first sound and

$$u_2 = \sqrt{\frac{\rho_s \sigma^2}{\rho_n\left(\frac{\partial\sigma}{\partial T}\right)}}$$

the velocity of second sound (cf. Eqs. 10-14 and 10-15).

Thus in narrow capillaries there can exist soundlike oscillations of the superfluid component whose velocity is given by Eq. 14-7. These oscillations were named fourth sound by Atkins. In practice in formula 14-7 the second term turns out to be much smaller than the first at all temperatures.

15

CAPILLARY WAVES

There can exist waves that propagate along a free surface of helium II, but are damped in the direction going into the liquid. This phenomenon is completely analogous to capillary waves on the surface of a classical liquid. Let us choose the z-axis along the normal to the surface, and denote by ζ the deviation of the coordinates of points on the surface from their equilibrium values. If effects linked to the presence of vapor are neglected, then at the surface of the liquid the following boundary conditions should be satisfied

(a) The formal flux of liquid across the surface should be equal to zero

$$\rho_s v_{sz} + \rho_n v_{nz} - \rho\dot{\zeta} = 0 \tag{15-1}$$

(b) The sum of the pressure forces and the surface tension should vanish (the surface is in the xy-plane)

$$p - \gamma\left(\frac{\partial^2 \zeta}{\partial x^2} + \frac{\partial^2 \zeta}{\partial y^2}\right) = 0 \tag{15-2}$$

(c) The entropy flux across the surface should vanish

$$v_{nz} - \dot{\zeta} = 0 \tag{15-3}$$

From Eqs. 15-1 and 15-3 it follows that z-components of the normal and superfluid velocities should be equal to the velocity of the surface $\dot{\zeta}$.

Let us consider a periodic wave propagating in the x-direction and decaying in the inward direction. In such a wave all quantities will vary as $\exp(-i\omega t + ikx - \kappa z)$. Since the linearized hydrodynamic equations become wave equations the frequency ω, the wave vector k, and the quantity κ are related by the equation

$$\frac{\omega^2}{u^2} = k^2 - \kappa^2 \qquad (15\text{-}4)$$

In the limit of low frequencies we have simply

$$k = \kappa$$

The z-component of the equation of motion gives a relation which should hold at the surface

$$-i\omega j_z' = \kappa p' \qquad (15\text{-}5)$$

(we denote by a prime the varying part of any quantity). From condition 15-2 we have

$$p' + \gamma k^2 \zeta = 0 \qquad (15\text{-}6)$$

Since $j_z' = \rho\dot{\zeta}$ we finally obtain from Eq. 15-5 and 15-6

$$\omega^2 = \frac{\gamma}{\rho} k^3 \qquad (15\text{-}7)$$

Thus capillary waves have the same dispersion law in a superfluid as in a classical liquid.

The situation changes if we consider waves propagating along a helium II film. In this case, if the thickness of the film is very small, the normal component does not participate in the oscillatory motion and the boundary conditions are changed.

Instead of Eq. 15-1 we have

$$\rho_s v_{sz} - \rho\dot{\zeta} = 0 \qquad (15\text{-}8)$$

The equation of motion of the superfluid part yields

$$i\omega v_{sz} + \kappa\mu' = 0 \qquad (15\text{-}9)$$

From Eqs. 15-8, 15-9, and 15-6 we obtain a new dispersion equation

$$\omega^2 = \frac{\rho_s}{\rho}\frac{\gamma}{\rho} k^3 \qquad (15\text{-}10)$$

This equation holds only if the wavelength of the waves is much smaller than the thickness of the film h ($kh \gg 1$). In the general case a similar calculation gives

$$\omega^2 = \frac{\rho_s}{\rho}\frac{\gamma}{\rho} k^3 \cdot \tanh kh \tag{15-11}$$

Such waves, which propagate in the superfluid component ($\mathbf{v}_n = 0$) along a film, are sometimes called third sound.

16

THE HYDRODYNAMICS OF A ROTATING SUPERFLUID[19]

The basis of the hydrodynamic analysis of superfluid helium carried out in Chap. 8 was the experimentally verified assertion that the superfluid motion is irrotational. It follows immediately from this that if we rotate a cylindrical bucket of superfluid helium, only the normal component should be carried along by the rotation. The superfluid part should remain stationary. This also follows from our simple microscopic picture. Indeed, when the bucket rotates the excitations collide with the walls and are carried along by the walls, so that the normal component of the liquid moves with the bucket. The superfluid component does not interact with the walls and remains stationary. However this conclusion is not confirmed experimentally. If the above picture were accurate, then the height of the meniscus formed in a rotating bucket of helium II would be smaller by a factor ρ_n/ρ than the one formed in a classical liquid. But this is not what is observed experimentally. Rotating helium II forms a meniscus that has exactly the same height as that of a classical liquid. This means that experiment shows that when superfluid helium is rotated, the *whole* liquid is set into motion.

In order to understand this situation let us return once more to the condition stating that superfluid motion is irrotational. This is expressed by the relation curl $\mathbf{v}_s = 0$. One may rewrite this last condition using Stoke's theorem to state that the circulation of the superfluid velocity along any contour vanishes

$$\oint_L \mathbf{v}_s \cdot d\boldsymbol{\ell} = 0 \qquad (16\text{-}1)$$

Let us multiply relation 16-1 by the mass m of a helium atom. In this manner we obtain on the left a quantity which in quantum mechanics usually obeys a quantization condition. It is therefore natural to consider 16-1 as a special case of a more general quantum condition

$$m \oint \mathbf{v}_s \cdot d\boldsymbol{\ell} = 2\pi n\hbar \qquad (16\text{-}2)$$

(n is an integer).

We may attempt to obtain this quantum condition on the circulation for a Bose liquid starting from first principles.[20]

Let us consider the liquid at zero temperature when it is in its ground state. The wave function ϕ_0 in this state is a symmetric function of all the coordinates and has no nodes. The function ϕ_0 can be chosen as real and in what follows can be considered to be merely a positive number. Let the whole system move with the uniform velocity $\mathbf{v}_s$. The wave function will then have a phase $(i/\hbar)Nm\mathbf{v}_s \cdot \Sigma \mathbf{r}_i/N$ and will be written in the form (the sum extends over all N particles of the system)

$$\phi = \phi_0 \exp\left(\frac{i}{\hbar} m\mathbf{v}_s \cdot \sum_i \mathbf{r}_i\right) \qquad (16\text{-}3)$$

Since the wave function has no nodes we may conclude that it is a continuous function of the phase. Formula 16-3 is exact for uniform motion. In the general case, when the velocity $\mathbf{v}_s$ varies from point to point, we may introduce the change in phase $\Delta\varphi$ caused by a small change $\Delta\mathbf{r}_i$ in the coordinates of all the particles

$$\frac{i}{\hbar} m \sum \mathbf{v}_s \cdot \Delta\mathbf{r}_i \qquad (16\text{-}4)$$

Let us choose in our system a certain set of particles that form a closed contour. The displacement of the particles along the contour does not change the state of the system, so that the phase of the wave function can only change by an integer multiple of 2π. In the limit of infinitesimal $\Delta\mathbf{r}_i$ we therefore obtain

$$\frac{m}{\hbar} \sum \mathbf{v}_s \cdot \Delta\mathbf{r}_i \rightarrow \frac{m}{\hbar} \oint \mathbf{v}_s \cdot d\boldsymbol{\ell} = 2\pi n \qquad (16\text{-}5)$$

which is just the general circulation theorem we were looking for.

At nonzero temperatures, each one of the excitations present gives a contribution

$$\sum \exp\left(\frac{i}{\hbar} \mathbf{p} \cdot \mathbf{r}_i\right) \qquad (16\text{-}6)$$

to the phase of the wave function ($\mathbf{p}$ is the momentum of the excitation).

But the part of the phase which corresponds to motion of the system with velocity $\mathbf{v}_s$ is unchanged. If in Eq. 16-5 we choose the contour L in such a way that it does not go near any excitation, then the whole previous analysis remains unchanged.

From Eq. 16-2 we see that in a superfluid two distinct situations are possible depending on whether n is equal to zero or not. For $n = 0$ we have $\text{curl}\ \mathbf{v}_s = 0$ and for rotation in a simply connected region it follows that $\mathbf{v}_s = 0$. For $n \neq 0$ the situation is more complicated. The circulation around certain singular lines is then not equal to zero. On these lines, which are analogous to vortex lines in classical hydrodynamics, v_s has a singularity. Clearly near the vortex lines the above arguments are not valid. However if we are not interested in the detailed structure of the vortex core, the only limitation on our arguments is that the contour L should not come too close to this core.

Let us now investigate the motion of the superfluid in a rotating bucket. The velocity field per unit length of a vortex parallel to the axis of rotation is, according to Eq. 16-2, equal to

$$v_s = \frac{\kappa}{2\pi} \times \frac{1}{r} \qquad \kappa = \frac{h}{m} n \tag{16-7}$$

The kinetic energy per unit length is

$$E_k = \tfrac{1}{2} \rho_s \int_a^b v_s^2\, 2\pi r\, dz = \rho_s \frac{\kappa^2}{4\pi} \ln \frac{b}{a} \tag{16-8}$$

where a is the radius of the vortex core, of the order of the interatomic distance, and b is some external dimension of the vortex. This dimension is connected with the total number of vortices N per unit surface by the obvious relation

$$\pi b^2 = \frac{1}{N}$$

Finally we may write down the connection between the density of vortices N and the quantity $\text{curl}\ \mathbf{v}_s$. According to Eq. 16-2 we have, using Stoke's theorem,

$$N = \frac{|\text{curl}\ \mathbf{v}_s|}{\kappa} \tag{16-9}$$

In order to study the rotation of a superfluid let us use a variational method. In equilibrium, the quantity

$$F - M\omega_0$$

has a minimum; F is the free energy, M the angular momentum of the liquid, ω_0 the angular velocity of rotation of the bucket. The free energy of the moving liquid is made up of the free energy of the liquid at rest, plus the kinetic energy of motion†

$$\rho_s \frac{v_s^2}{2} + \rho_n \frac{v_n^2}{2}$$

Since the kinetic energy and the angular momentum are additive, we may consider the motion of the superfluid component independently of the normal motion and minimize the quantity

$$E_s - M_s \omega_0$$

where E_s and M_s refer to the superfluid part. As is easy to see the normal part will undergo a rotation with velocity $v_n = \omega_0 r$, just like a solid body. The energy E of superfluid motion (in what follows we shall omit the index s for simplicity) is equal to

$$E = \tfrac{1}{2} \rho_s \int v^2 \, 2\pi r \, dr + \rho_s \frac{\kappa}{4\pi} \times \int |\text{curl } \mathbf{v}| \ln\left(\frac{\kappa^{1/2}}{\pi^{1/2} |\text{curl } \mathbf{v}|^{1/2} a}\right) 2\pi r \, dr \qquad (16\text{-}10)$$

The second term is the energy of the vortices calculated with the aid of Eqs. 16-7 to 16-9. In a similar manner we may obtain the angular momentum of the liquid.

$$M = \rho_s \int v r 2\pi r \, dr + \rho_s \frac{\kappa}{2\pi} \int 2\pi r \, dr \qquad (16\text{-}11)$$

Here the second term is due to the vortices, just as in Eq. 16-10. Since the calculation of the vortex contribution was only performed with logarithmic accuracy, the inclusion of this term is strictly speaking beyond the range of accuracy of our computation and we could have left it out. Let us vary the difference $E - M\omega_0$ with respect‡ to δv

---

†Strictly speaking such a separation is approximate, since the thermodynamic variables depend on the relative velocity $v_n - v_s$. However, to the accuracy with which we are calculating, this fact is not important in this problem.

‡We remind the reader that in cylindrical coordinates

$$|\text{curl } \mathbf{v}| = \frac{1}{r} \frac{\partial}{\partial r}(vr)$$

$$\int \delta v \left\{ (v - \omega_0 r) + \frac{\kappa}{8\pi} \frac{\frac{\partial}{\partial r}\frac{1}{r}\frac{\partial}{\partial r}(vr)}{\frac{1}{r}\frac{\partial}{\partial r}(vr)} \right\} 2\pi r \, dr$$

$$+ \frac{\kappa}{4\pi} \int \frac{\partial}{\partial r} \left\{ r \, \delta v \ln \frac{\kappa^{1/2}}{(\pi e \, | \operatorname{rot} v |)^{1/2} a} \right\} 2\pi \, dr \qquad (16\text{-}12)$$

By requiring that the first integral should vanish we obtain the equation

$$(v - \omega_0 r) \frac{1}{r} \frac{\partial}{\partial r}(vr) + \frac{\kappa}{4\pi} \frac{\partial}{\partial r} \frac{1}{r} \frac{\partial}{\partial r}(vr) = 0 \qquad (16\text{-}13)$$

The second integral in Eq. 16-12 will automatically vanish if δv is equal to zero at the boundaries of the domain of integration.

Equation 16-13 determines the superfluid velocity in a rotating cylindrical bucket. It has two exact solutions

$$v = \omega_0 r \qquad (16\text{-}14)$$

$$v = \frac{K}{r} \qquad (16\text{-}15)$$

The first solution corresponds to solid body rotation and the second to irrotational motion with curl $\mathbf{v} = 0$. It should be noted that the sum of solutions Eqs. 16-14 and 16-15 does not satisfy Eq. 16-13. Let us suppose that solid body rotation occurs in a region inside some radius R_i. Outside this region we have irrotational motion given by Eq. 16-15. If we minimize the free energy with respect to the radius R_i we find a value of R_i which as we shall see is less than the radius R of the bucket. Thus a calculation confirms our assumption. The constant K is determined from the boundary condition $v(R) = \omega_0 R$.† The solutions obtained by using this boundary condition does not match at $r = R_i$ with the solution, Eq. 16-14. According to Eq. 16-13 the quantities vr and $(1/r)(\partial/\partial r)(vr)$ should be continuous. Therefore for values of r close to R_i the solutions 16-14 and 16-15 should in any case be corrected. It is easy to see that the only parameter in Eq. 16-13 having the dimensions of length is $\sqrt{\kappa/\omega_0}$. The true solution will only differ from the one we have chosen in a narrow region around R_i of order $\sqrt{\kappa/\omega_0}$.‡ The region of irrotational motion has dimensions of order $\sqrt{(\kappa/\omega_0)} \ln (b/a)$, as we shall see below. Since

---

†It is essential to insert this boundary condition for v, since we varied the free energy under the assumption that at the boundaries $\delta v = 0$.

‡This is confirmed by more detailed calculations.

our calculations have only logarithmic accuracy, we may simply neglect the small region around R_i where we should correct solutions 16-14 and 16-15, and consider that inside R_i we have the velocity field $v = \omega_0 r$ and outside the field $v = \omega_0 R^2/r$. Let us insert this velocity field into the expression for $E - M\omega_0$ and minimize it with respect to R_i. We thus obtain

$$\left(\frac{R}{R_i} - \frac{R_i}{R}\right)^2 = \frac{\kappa}{\omega_0 R^2} \ln \frac{b}{ae} \tag{16-16}$$

The right-hand side of Eq. 16-16 is small in all interesting cases so that we have approximately

$$R - R_i = \frac{1}{2}\sqrt{\frac{\kappa}{\omega_0} \ln \frac{b}{ae}} \tag{16-17}$$

From this formula we see that the region of irrotational motion is comparatively small. However, since the logarithm is rather large (of order 15) this region is observable. There will also be regions of irrotational motion in the case of rotation of helium between coaxial cylinders. In this case there will be two such regions, near the inner and near the outer boundaries.

Let us now calculate the free energy $E - M\omega_0$ of the superfluid component. Inserting the value of R_i into Eqs. 16-10 and 16-11 and taking into account the smallness of $R - R_i$ we find

$$E - M\omega_0 = -\tfrac{1}{4}\pi\rho_s \omega_0 R^2 + \tfrac{1}{2}\rho_s \kappa\omega_0 R^2 \ln \frac{b}{ae} \tag{16-18}$$

The second term in 16-18 is due to the presence of vortices, and, as we can see, it increases the free energy. Therefore the vortices that arise will have the smallest possible angular momentum, corresponding to the value $n = 1$.

We thus see that when we rotate a bucket containing superfluid helium, the vortices that arise imitate vortex motion in practically the whole bucket, with a velocity equal to twice the velocity of rotation of the bucket. That is, they behave just like a solid body or a classical viscous fluid. However, in the whole volume of the liquid which is not occupied by vortices, we have curl $\mathbf{v}_s = 0$. Only a small region near the walls of the bucket is completely free from vortices; in this region the motion is completely irrotational.

CONCERNING CRITICAL VELOCITIES

The critical velocities for creation of rotons or phonons, which according to the Landau criterion are equal to $v_{cr} = \min \epsilon/p$, are two orders of magnitude larger than the experimentally observed values and do not depend on the width of the capillary. The experimental values of the critical velocities and the correct dependence on the width of the capillary immediately follow from the possibility of creation of vortex rings in the superfluid. These curious excitations are characterized by an energy

$$\epsilon = \rho_s \frac{Rh^2}{2m^2} \ln \frac{R}{a} \tag{16-19}$$

and a momentum

$$p = \rho \frac{\pi R^2 h}{m} \tag{16-20}$$

which depend on the radius of the ring R (a is the radius of the vortex core).

Applying the Landau criterion to this branch of the spectrum we obtain

$$v_{cr} = \min \frac{\epsilon}{p} = \frac{\hbar}{Rm} \ln \frac{R}{a} \tag{16-21}$$

This formula explains the observed fact that $v_{cr} R$ is constant and gives values of the critical velocities which are close to the observed ones.

THE HYDRODYNAMIC EQUATIONS OF ROTATING SUPERFLUIDS

The hydrodynamic equations of a superfluid in the presence of vortices can be obtained phenomenologically by using the conservation laws. In this manner the form of the dissipative terms connected with the existence of a nonvanishing curl $\mathbf{v}_s$ also becomes clear. The primary difference between this situation and the case of irrotational superfluid motion, is expressed by the dependence of the internal energy of the liquid on the absolute value of curl $\mathbf{v}_s$. This dependence can be expressed in a differential form as

$$\delta\epsilon = \lambda\ \delta\omega \qquad \omega = |\text{curl } \mathbf{v}_s| \tag{16-22}$$

According to Eq. 16-8 the coefficient λ is, to within logarithmic accuracy, equal to

$$\lambda = \rho_s \frac{\kappa}{4\pi} \ln \frac{b}{a} \tag{16-23}$$

From here on we may carry out the analysis in the same way as in Chap. 8, where we derived the hydrodynamic equations for curl $\mathbf{v}_s = 0$. Let us write all equations in the form of differential conservation laws. The equation of continuity remains unchanged.

$$\dot{\rho} + \operatorname{div} \mathbf{j} = 0 \tag{16-24}$$

and, in fact, is just the definition of the mass current vector $\mathbf{j}$ of the liquid. We add to the momentum flux tensor Π_{ik} and to the energy flux vector $\mathbf{Q}$ the terms π_{ik} and $\mathbf{q}$ which for the moment remain unknown. In this manner the energy and momentum conservation equations are written in the form

$$\frac{\partial E}{\partial t} + \operatorname{div} (\mathbf{Q} + \mathbf{q}) = 0 \tag{16-25}$$

$$\frac{\partial j_i}{\partial t} + \frac{\partial}{\partial r_k} (\Pi_{ik} + \pi_{ik}) \tag{16-26}$$

By making a Galilean transformation we express the energy per unit volume E in terms of the internal energy E_0 in the frame of reference moving with the superfluid velocity

$$E = \tfrac{1}{2} \rho v_s^2 + \mathbf{j}_0 \cdot \mathbf{v}_s + E_0 \tag{16-27}$$

where $\mathbf{j}_0$ is the momentum per unit volume in the same system, which is expressible in terms of the total momentum in the rest frame by the relation

$$\mathbf{j}_0 = \mathbf{j} - \rho \mathbf{v}_s \tag{16-28}$$

In the thermodynamic identity defining the internal energy E_0 it is important to add to the terms contained in Eq. 8-12 the change in energy due to the presence of curl $\mathbf{v}_s$. We thus have

$$dE_0 = T\, dS + \mu\, d\rho + ((\mathbf{v}_n - \mathbf{v}_s) \cdot d\mathbf{j}_0) + \lambda\, d\omega \tag{16-29}$$

The unperturbed momentum flux tensor and energy density vector are given by Eqs. 8-19 and 8-20

$$\Pi_{ik} = \rho v_{si} v_{sk} + v_{si} j_{0k} + v_{nk} j_{0i} + p\, \delta_{ik}$$

$$\mathbf{Q} = \left(\mu + \frac{\mathbf{v}_s^2}{2}\right) \mathbf{j} + ST\mathbf{v}_n + \mathbf{v}_n (\mathbf{v}_n \cdot \mathbf{j}_0)$$

where the pressure p is equal to

$$p = -E_0 + TS + \mu\rho + ((\mathbf{v}_n - \mathbf{v}_s)\cdot\mathbf{j}_0)$$

The complete system of hydrodynamic equations contains still two equations, the equation of superfluid motion and the equation of increase of entropy

$$\frac{\partial \mathbf{v}_s}{\partial t} + (\mathbf{v}_s\cdot\nabla)\mathbf{v}_s + \nabla\mu = \mathbf{f} \tag{16-30}$$

$$\frac{\partial S}{\partial t} + \operatorname{div} S\mathbf{v}_n = \frac{R}{T} \tag{16-31}$$

where the quantity $\mathbf{f}$ and the dissipative function are as yet unknown. Let us now use the law of conservation of energy, Eq. 16-25, in order to determine the additional terms in the hydrodynamic equations. For this, let us take the time derivative of Eq. 16-27 and insert from Eqs. 16-24, 16-26, and 16-30 the time derivatives of the quantities $\mathbf{j}$, ρ, and $\mathbf{v}_s$. As a result we obtain

$$\dot{E} = -\operatorname{div}\mathbf{Q} - \operatorname{div}(\pi\cdot\mathbf{v}_n) + T(\dot{S} + \operatorname{div} S\mathbf{v}_n) + \lambda\dot{\omega} + \lambda v_{ni}\frac{\partial\omega}{\partial r_i}$$
$$+ \pi_{ik}\frac{\partial v_{ni}}{\partial r_k} + (\mathbf{j} - \rho\mathbf{v}_n)\cdot[\mathbf{f} + \omega\times(\mathbf{v}_n - \mathbf{v}_s)] \tag{16-32}$$

where

$$(\pi\cdot\mathbf{v}_n)_i = \pi_{ik}v_{nk}$$

Let us introduce the unit vector $\boldsymbol{\nu} = \boldsymbol{\omega}/\omega$ and calculate the derivative $\dot{\omega}$ with the aid of Eq. 16-30

$$\dot{\omega} = \boldsymbol{\nu}\cdot\operatorname{curl}\mathbf{v}_s$$
$$= \boldsymbol{\nu}\cdot\operatorname{curl}[\mathbf{f} + \boldsymbol{\omega}\times(\mathbf{v}_n - \mathbf{v}_s)] - \boldsymbol{\nu}\cdot\operatorname{curl}(\boldsymbol{\omega}\times\mathbf{v}_n) \tag{16-33}$$

We insert this expression into Eq. 16-32 and group all the terms which have the form of a divergence, obtaining

$$\dot{E} + \operatorname{div}\{\mathbf{Q} + (\pi\cdot\mathbf{v}_n) + \lambda\boldsymbol{\nu}\times[\mathbf{f} + \boldsymbol{\omega}\times(\mathbf{v}_n - \mathbf{v}_s)]\}$$
$$= T(\dot{S} + \operatorname{div} S\mathbf{v}_n) + \left(\pi_{ik} + \lambda\frac{\omega_i\omega_k}{\omega} - \lambda\omega\,\delta_{ik}\right)\frac{\partial v_{ni}}{\partial r_k}$$
$$+ [\mathbf{f} + \boldsymbol{\omega}\times(\mathbf{v}_n - \mathbf{v}_s)]\cdot[\mathbf{j} - \rho\mathbf{v}_n + \operatorname{curl}\lambda\boldsymbol{\nu}] \tag{16-34}$$

Comparing Eqs. 16-34 with the law of conservation of energy, Eq. 16-25, and the law of increase of entropy, Eq. 16-31, we have

$$\mathbf{q} = (\pi \cdot \mathbf{v}_n) + \lambda\boldsymbol{\nu} \times [\mathbf{f} + \boldsymbol{\omega} \times (\mathbf{v}_n - \mathbf{v}_s)] \tag{16-35}$$

$$R = -\left(\pi_{ik} + \lambda \frac{\omega_i \omega_k}{\omega} - \lambda\omega\, \delta_{ik}\right) \frac{\partial v_{ni}}{\partial r_k}$$

$$- [\mathbf{f} + \boldsymbol{\omega} \times (\mathbf{v}_n - \mathbf{v}_s)] \cdot [\mathbf{j} - \rho\mathbf{v}_n + \operatorname{curl} \lambda\boldsymbol{\nu}] \tag{16-36}$$

From the requirement that the dissipative function R should be positive we find the most general form for the function **f** and the tensor π_{ik}

$$\mathbf{f} = -\boldsymbol{\omega} \times (\mathbf{v}_n - \mathbf{v}_s) + \alpha\boldsymbol{\omega} \times [\mathbf{j} - \rho\mathbf{v}_n + \operatorname{curl} \lambda\boldsymbol{\nu}]$$

$$+ \beta\boldsymbol{\nu} \times \{\boldsymbol{\omega} \times [\mathbf{j} - \rho\mathbf{v}_n + \operatorname{curl} \lambda\boldsymbol{\nu}]\}$$

$$- \gamma\boldsymbol{\nu}[\boldsymbol{\omega} \cdot (\mathbf{j} - \rho\mathbf{v}_n + \operatorname{curl} \lambda\boldsymbol{\nu})] \qquad \beta, \gamma \geq 0 \tag{16-37}$$

$$\pi_{ik} = \lambda\omega\, \delta_{ik} - \lambda \frac{\omega_i \omega_k}{\omega} + \tau_{ik} \tag{16-38}$$

The viscous stress tensor τ_{ik} may as usual be expressed in terms of the viscosity coefficients (cf. Eq. 9-9).

According to Eq. 16-28 the momentum of relative motion is equal to†

$$\mathbf{j}_0 = \rho_n (\mathbf{v}_n - \mathbf{v}_s)$$

Expressing the difference $\mathbf{j} - \rho\mathbf{v}_n$ in terms of $\mathbf{v}_n - \mathbf{v}_s$

$$\mathbf{j} - \rho\mathbf{v}_n = \mathbf{j}_0 + \rho\mathbf{v}_s - \rho\mathbf{v}_n = -\rho_s (\mathbf{v}_n - \mathbf{v}_s)$$

we transform the expression for **f** into

$$\mathbf{f} = -\frac{1}{\rho_s}\, \boldsymbol{\omega} \times \operatorname{curl} \lambda\boldsymbol{\nu} - (1 + \alpha\rho_s)\, \boldsymbol{\omega} \times \left(\mathbf{v}_n - \mathbf{v}_s - \frac{1}{\rho} \operatorname{curl} \lambda\boldsymbol{\nu}\right)$$

$$- \beta\rho_s\, \boldsymbol{\nu} \times \boldsymbol{\omega} \times \left(\mathbf{v}_n - \mathbf{v}_s - \frac{1}{\rho_s} \operatorname{curl} \lambda\boldsymbol{\nu}\right)$$

$$+ \gamma\rho_s\, \boldsymbol{\nu} \left[\boldsymbol{\omega} \cdot \left(\mathbf{v}_n - \mathbf{v}_s - \frac{1}{\rho_s} \operatorname{curl} \lambda\boldsymbol{\nu}\right)\right] \tag{16-39}$$

---

†Since the direction $\boldsymbol{\omega}$ is singled out in our liquid, there may in general be effects due to the anisotropy of the liquid. As a consequence, instead of the viscosity η there arises a viscosity tensor $\eta_{ik\ell m}$, and in the momentum $\mathbf{j}_0$ there appears a component parallel to the direction $\boldsymbol{\omega}$, etc. However these effects are quadratic in κ and are extremely small.

Let us write the equation of superfluid flow, Eq. 16-30, in its final form

$$\frac{\partial \mathbf{v}_s}{\partial t} + (\mathbf{v}_s \cdot \nabla)\mathbf{v}_s + \nabla\mu$$

$$= \frac{1}{\rho_s}\,\boldsymbol{\omega} \times \operatorname{curl}\lambda\boldsymbol{\nu} - \beta'\rho_s\,\boldsymbol{\omega} \times \left(\mathbf{v}_n - \mathbf{v}_s - \frac{1}{\rho_s}\operatorname{curl}\lambda\boldsymbol{\nu}\right)$$

$$- \beta\rho_s\,\boldsymbol{\nu} \times \left[\boldsymbol{\omega} \times \left(\mathbf{v}_n - \mathbf{v}_s - \frac{1}{\rho_s}\operatorname{curl}\lambda\boldsymbol{\nu}\right)\right]$$

$$+ \gamma\rho_s\,\boldsymbol{\nu}\left[\boldsymbol{\omega} \cdot \left(\mathbf{v}_n - \mathbf{v}_s - \frac{1}{\rho_s}\operatorname{curl}\lambda\boldsymbol{\nu}\right)\right] \qquad (16\text{-}40)$$

Here $\beta'\rho_s = 1 + \alpha\rho_s$. The last three terms describe the force of mutual friction, the coefficients β and β' being expressible in terms of the quantities B and B′ introduced by Hall and Vinen,[21] by the relations

$$\beta = \tfrac{1}{2}B\,\frac{\rho_n}{\rho\rho_s} \qquad \beta' = \tfrac{1}{2}B'\,\frac{\rho_n}{\rho\rho_s} \qquad (16\text{-}41)$$

In the case of transverse deflections of the vortices the mutual friction contains additional terms in curl $\lambda\boldsymbol{\nu}$. The longitudinal force (the last term in Eq. 16-40) acts if the direction of individual vortices is different from the average direction of the vortices, for example, in the case of thermal oscillations. In view of the smallness of this effect the coefficient γ appears to be extremely small. The addition to the momentum flux tensor caused by the vortex motion

$$\tilde{\pi}_{ik} = \lambda\omega\,\delta_{ik} - \lambda\,\frac{\omega_i\,\omega_k}{\omega} \qquad (16\text{-}42)$$

consists of two parts: a renormalization of the pressure $\lambda\omega$ and a term which may be interpreted as the vortex filament tension $-\lambda\omega_i\,\omega_k/\omega$. From Eq. 16-35, by inserting the expression for $\tilde{\pi}_{ik}$ given in Eq. 16-42, we find the additional energy flux $\mathbf{q}$

$$\mathbf{q} = \lambda\boldsymbol{\nu} \times [\mathbf{f} + (\mathbf{v}_s \times \boldsymbol{\omega})] \qquad (16\text{-}43)$$

In the absence of a longitudinal force $(\gamma = 0)$ we may transform this expression into a form which permits a simple interpretation. Let us take the curl of Eq. 16-40,

$$\frac{\partial \boldsymbol{\omega}}{\partial t} = \operatorname{curl}\{\mathbf{f} + [\mathbf{v}_s \times \boldsymbol{\omega}]\} = \operatorname{curl}(\mathbf{v}_L \times \boldsymbol{\omega}) \qquad (16\text{-}44)$$

This equation describes the transport of the vector ω with a velocity

$$\mathbf{v}_L = \mathbf{v}_s - \frac{1}{\rho_s}\operatorname{curl}\lambda\boldsymbol{\nu} + \beta'\rho_s\left(\mathbf{v}_n - \mathbf{v}_s - \frac{1}{\rho_s}\operatorname{curl}\lambda\boldsymbol{\nu}\right) - \beta\rho_s\boldsymbol{\nu}\times\left(\mathbf{v}_n - \mathbf{v}_s - \frac{1}{\rho_s}\operatorname{curl}\lambda\boldsymbol{\nu}\right) \qquad (16\text{-}45)$$

This is the velocity of motion of the vortex filaments (in the absence of longitudinal friction). The energy flux $\mathbf{q}$ may now be written in the final form

$$\mathbf{q} = \omega\frac{\partial E_0}{\partial\omega}\boldsymbol{\nu}\times(\mathbf{v}_L\times\boldsymbol{\nu}) \qquad (16\text{-}46)$$

from which it is clear that the energy of the vortices is transported in a direction perpendicular to ω.

Let us now consider the boundary conditions satisfied by the superfluid component at the surface of a solid body and at the free surface, as the normal component of the velocity $\mathbf{v}_s$ goes to zero.

In the case of a free surface, the tangential components of the tension in the liquid must as usual be equal—that is, $\pi_{\mathfrak{N}\tau} = \omega_{\mathfrak{N}}\omega_\tau = 0$. Consequently the vortices are perpendicular to the surface ($\mathfrak{N}$ is the normal to the surface). At a solid boundary the situation is somewhat more complicated. If the surface is rough the vortices move with the surface. In the case of vortex slippage on the surface, energy dissipation will take place. By calculating the energy dissipation it is possible to show, in the general case of a surface moving with velocity $\mathbf{u}$, that if $\gamma = 0$, we have

$$\mathbf{v}_L - \mathbf{u} = \zeta\boldsymbol{\nu}\times(\mathfrak{N}\times\boldsymbol{\nu}) + \zeta'\mathfrak{N}\times\boldsymbol{\nu} \qquad (16\text{-}47)$$

where the coefficients ζ and ζ' are in order of magnitude

$$\frac{\zeta}{B} \approx \frac{\zeta'}{B'} \approx \frac{\rho_n\hbar}{\rho m d} \qquad (16\text{-}48)$$

(d is the average dimension of the surface irregularity).

An absolutely rough surface corresponds to $\zeta = \zeta' = 0$ and an absolutely smooth surface to ζ and $\zeta' \to \infty$.

17

SUPERFLUID HYDRODYNAMICS NEAR THE λ-POINT[22]

In the theory of second-order phase transitions one usually introduces a small parameter, describing how close the state of the system is to the λ-point. The thermodynamic potentials are then expanded in a series in this parameter whose value is determined by minimizing the potential. Far from the λ-point we are dealing with an ideal gas of excitations and the normal density ρ_n (and consequently also the superfluid density ρ_s) can be calculated by using formula 2-22. Near the λ-point such an approach is not possible and one must use a method of expansion in a small parameter. The density ρ_s may be used as such a parameter, since it vanishes at the λ-point. It turns out to be more convenient to introduce a complex function $\psi(x,y,z,t) = \eta e^{i\varphi}$ defined in such a way that

$$\rho_s = m\,|\psi|^2 \qquad \mathbf{v}_s = \frac{\hbar}{m}\nabla\varphi \tag{17-1}$$

For small enough values of the velocities $\mathbf{v}_n$ and $\mathbf{v}_s$ the energy per unit volume E may be expanded in a series in $\mathbf{v}_n$ and $\nabla\psi$

$$E = (\rho - m\,|\psi|^2)\frac{\mathbf{v}_n^2}{2} + \frac{\hbar^2}{2m}\,|\nabla\psi|^2 + E_0(\rho, S, |\psi|^2) \tag{17-2}$$

If we further express ψ in terms of ρ_s and $\mathbf{v}_s$ we obtain

$$E = \rho_n\frac{\mathbf{v}_n^2}{2} + \rho_s\frac{\mathbf{v}_s^2}{2} + E_0 + \frac{\hbar^2}{8m^2}\,\frac{(\nabla\rho_s)^2}{\rho_s} \tag{17-3}$$

$$\rho_n = \rho - m\,|\psi|^2$$

The last term in Eq. 17-3 represents a specifically quantum-mechanical contribution.

The equilibrium values of ψ and ψ^* are found by minimizing the energy E with respect to ψ^* and ψ for a given value of the momentum $\mathbf{j}$

$$\mathbf{j} = \rho_n \mathbf{v}_n + \rho_s \mathbf{v}_s$$

$$= (\rho - m|\psi|^2)\mathbf{v}_n + \frac{i\hbar}{2}(\psi\nabla\psi^* - \psi^*\nabla\psi) \qquad (17\text{-}4)$$

Let us multiply Eq. 17-4 by an undetermined Lagrange multiplier $\mathbf{u}$ and add to Eq. 17-3. We then vary the sum with respect to ψ^* and v_n (at constant ρ, s, j, and ψ) and eliminate $\mathbf{u}$. We thus find the equation

$$\frac{1}{2}\left(-\frac{i\hbar}{m}\nabla - \mathbf{v}_n\right)^2 \psi + \left(\frac{\partial E_0}{\partial \rho_s}\right)_{\rho,S} \psi = 0 \qquad (17\text{-}5)$$

which determines the equilibrium value of ψ.

Let us now derive the complete system of hydrodynamic equations. These may be obtained by starting from the conservation laws as was done previously in the derivation of the two-fluid hydrodynamics. We start with the equation determining the function ψ. This may be found from the assumption that the state of the system is determined by fixing ψ (just as for other thermodynamic quantities)—that is, ψ satisfies a linear differential equation in t. In analogy with quantum mechanics we may write

$$i\hbar \frac{\partial \psi}{\partial t} = \hat{L}\psi \qquad (17\text{-}6)$$

where $\hat{L}$ is some linear operator. Since the quantity $\rho_s = m|\psi|^2$ can relax (that is, it is not conserved but may be transformed into ρ_n), the operator $\hat{L}$ is not hermitian. The hermitian part of $\hat{L}$ can, in analogy with the Schrödinger equation, be represented in the form

$$-\frac{\hbar^2}{2m}\nabla^2 + V \qquad (17\text{-}7)$$

where

$$V = \left(\frac{\partial E_0}{\partial |\psi|^2}\right)_{\rho_n,S} - gm = \left\{\left(\frac{\partial E_0}{\partial \rho}\right)_{\rho_s} + \left(\frac{\partial E_0}{\partial \rho_s}\right) - g\right\} m \qquad (17\text{-}8)$$

and the function g is as yet undefined.

As for the anti-hermitian part of $\hat{L}$, it describes the approach of ρ_s to its equilibrium value and should vanish at equilibrium. If the departure of our state from equilibrium is small, then according to Eq. 17-5 it may be written in the form

$$-i\Lambda\left\{\frac{1}{2}\left(-\frac{i\hbar}{m}\nabla - \mathbf{v}_n\right)^2 m\psi + \frac{\partial E_0}{\partial \rho_s} m\psi\right\} \tag{17-9}$$

where Λ is some dimensionless coefficient proportional to the inverse relaxation time. The final equation for ψ is

$$i\hbar\frac{\partial\psi}{\partial t} = -\frac{\hbar^2}{2m}\nabla^2\psi + \left\{\left(\frac{\partial E_0}{\partial \rho}\right)_{\rho_s} + \left(\frac{\partial E_0}{\partial \rho_s}\right)_{\rho} - g\right\} m\psi$$

$$- i\Lambda\left\{\frac{1}{2}\left(-\frac{i\hbar}{m}\nabla - \mathbf{v}_n\right)^2 + \left(\frac{\partial E_0}{\partial \rho_s}\right)\right\} m\psi \tag{17-10}$$

As may be shown by a simple analysis the coefficient Λ must be real since ortherwise there would be a transport of the superfluid component with the normal velocity. All the other equations are written in their usual form:

~ the equation of continuity

$$\frac{\partial \rho}{\partial t} + \operatorname{div} \mathbf{j} = 0 \tag{17-11}$$

~ the law of conservation of momentum (cf. Eq. 8-3)

$$\frac{\partial j_i}{\partial t} + \frac{\partial \Pi_{ik}}{\partial r_k} = 0 \tag{17-12}$$

$$\Pi_{ik} = \frac{\hbar^2}{2m}\left(\frac{\partial\psi}{\partial r_i}\frac{\partial\psi^*}{\partial r_k} - \psi\frac{\partial^2\psi^*}{\partial r_i r_k} + \text{c.c.}\right)$$

$$+ \rho_n v_{ni} v_{nk} + p\,\delta_{ik} + \tau_{ik} \tag{17-13}$$

and finally the law of increase of entropy (cf. Eq. 9-16)

$$\frac{\partial S}{\partial t} + \operatorname{div}\left(S\mathbf{v}_n + \frac{\mathbf{q}}{T}\right) = \frac{R}{T} \tag{17-14}$$

The terms τ_{ik}, $\mathbf{q}$, and R, along with g, must still be defined. The law of conservation of energy requires that

$$\frac{\partial E}{\partial t} = -\operatorname{div} \mathbf{Q} \tag{17-15}$$

Let us further differentiate Eq. 17-2 with respect to time and express the time derivatives $\dot{\psi}$, $\dot{\rho}$, $\dot{S}$, and $\dot{\mathbf{v}}_n$ with the aid of Eqs. 17-11 to 17-14. Then let us choose the unknown terms in such a way that all terms which cannot be written as a divergence should vanish. Finally remembering that the function R is positive definite we find

$$\mathbf{q} = \kappa \nabla T$$

$$\tau_{ik} = -\eta\left(\frac{\partial v_{ni}}{\partial r_k} + \frac{\partial v_{nk}}{\partial r_i} - \frac{2}{3}\,\delta_{ik} \operatorname{div} \mathbf{v}_n\right) \tag{17-16}$$

As for the term g and the part of the tensor τ_{ik} with nonzero trace, we may set these equal to zero, since they are connected to the second viscosity, and near the λ-point the main contribution to dissipation comes from the relaxation of ρ_s, determined by the coefficient Λ.

Let us now write down the final equations obtained in this manner

$$i\hbar \frac{\partial \psi}{\partial t} = -\frac{\hbar^2}{2m}\nabla^2\psi + \left(\frac{\partial E_0}{\partial \rho} + \frac{\partial E_0}{\partial \rho_s}\right) m\psi$$
$$- i\Lambda\left\{\frac{1}{2}\left(-\frac{i\hbar}{m}\nabla - \mathbf{v}_n\right)^2 + \frac{\partial E_0}{\partial \rho_s}\right\} m\psi \tag{17-17}$$

$$\frac{\partial \rho}{\partial t} + \operatorname{div}\{(\rho - m|\psi|^2)\mathbf{v}_n\} + \frac{i\hbar}{2}(\psi\nabla^2\psi^* - \psi^*\nabla^2\psi) = 0 \tag{17-18}$$

$$\frac{\partial}{\partial t}\left\{(\rho - m|\psi|^2)v_{ni} + \frac{i\hbar}{2}\left(\psi\frac{\partial\psi^*}{\partial r_i} - \psi^*\frac{\partial\psi}{\partial r_i}\right)\right\} + \frac{\partial \Pi_{ik}}{\partial r_k} = 0 \tag{17-19}$$

$$\Pi_{ik} = (\rho - m|\psi|^2)v_{ni}v_{nk} + \frac{\hbar^2}{2m}\left(\frac{\partial\psi}{\partial r_i}\frac{\partial\psi^*}{\partial r_k} - \psi\frac{\partial^2\psi^*}{\partial r_i\,\partial r_k} + \text{c.c.}\right)$$
$$+ p\,\delta_{ik} - \eta\left(\frac{\partial v_{ni}}{\partial r_k} + \frac{\partial v_{nk}}{\partial r_i} - \frac{2}{3}\,\delta_{ik}\operatorname{div}\mathbf{v}_n\right) \tag{17-20}$$

$$p = -E_0 + TS + \rho\frac{\partial E_0}{\partial \rho} + \rho_s\frac{\partial E_0}{\partial \rho_s}$$

$$\frac{\partial S}{\partial t} + \operatorname{div} S\mathbf{v}_n$$
$$= \frac{1}{T}\operatorname{div}(\kappa \nabla T) + \frac{1}{T}\left\{\frac{2\Lambda}{\hbar}\left|\left[\frac{1}{2}\left(-\frac{i\hbar}{m}\nabla - \mathbf{v}_n\right)^2 + \frac{\partial E_0}{\partial \rho_s}\right] m\psi\right|^2\right.$$
$$\left. + \tfrac{1}{2}\eta\left(\frac{\partial v_{ni}}{\partial r_k} + \frac{\partial v_{nk}}{\partial r_i} - \frac{2}{3}\delta_{ik}\operatorname{div}\mathbf{v}_n\right)^2\right\} \qquad (17\text{-}21)$$

Let us consider the case of small gradients $\nabla\rho_s$, when the specifically quantum-mechanical effects are small. In this case we may everywhere replace $-(i\hbar/m)\nabla\psi$ by $\mathbf{v}_s$. If we take the gradient of Eq. 17-17 and make the aforementioned replacement in all the other equations we obtain a simpler system of equations

$$\dot{\mathbf{v}}_s + \nabla\left\{\frac{\mathbf{v}_s^2}{2} + \frac{\partial E_0}{\partial \rho} + \frac{\partial E_0}{\partial \rho_s} - \frac{\hbar\Lambda}{2m\rho_s}\operatorname{div}\rho_s(\mathbf{v}_s - \mathbf{v}_n)\right\} = 0 \qquad (17\text{-}22)$$

$$\frac{\partial \rho}{\partial t} + \operatorname{div}(\rho_s\mathbf{v}_s + \rho_n\mathbf{v}_n) = 0 \qquad (17\text{-}23)$$

$$\frac{\partial}{\partial t}(\rho_n v_{ni} + \rho_s v_{si}) + \frac{\partial}{\partial r_k}\left\{\rho_n v_{ni}v_{nk} + \rho_s v_{si}v_{sk} + p\,\delta_{ik}\right.$$
$$\left. - \eta\left(\frac{\partial v_{ni}}{\partial r_k} + \frac{\partial v_{nk}}{\partial r_i} - \frac{2}{3}\delta_{ik}\operatorname{div}\mathbf{v}_n\right)\right\} = 0 \qquad (17\text{-}24)$$

$$\frac{\partial \rho_s}{\partial t} + \operatorname{div}\rho_s\mathbf{v}_s = -\frac{\Lambda m}{2\hbar}\left\{\frac{(\mathbf{v}_n - \mathbf{v}_s)^2}{2} + \frac{\partial E_0}{\partial \rho_s}\right\}\rho_s \qquad (17\text{-}25)$$

$$\frac{\partial S}{\partial t} + \operatorname{div} S\mathbf{v}_n$$
$$= \frac{1}{T}\operatorname{div}(\kappa\nabla T) + \frac{1}{T}\left\{\frac{\hbar\Lambda}{2m\rho_s}[\operatorname{div}\rho_s(\mathbf{v}_s - \mathbf{v}_n)]^2\right.$$
$$+ \frac{2\Lambda m}{\hbar}\left[\tfrac{1}{2}(\mathbf{v}_s - \mathbf{v}_n)^2 + \left(\frac{\partial E_0}{\partial \rho_s}\right)\right]^2\rho_s$$
$$\left. + \frac{\eta}{2}\left(\frac{\partial v_{ni}}{\partial r_k} + \frac{\partial v_{nk}}{\partial r_i} - \frac{2}{3}\delta_{ik}\operatorname{div}\mathbf{v}_n\right)^2\right\} \qquad (17\text{-}26)$$

In these equations, in contrast to the hydrodynamic equations considered in Chap. 8, the quantity ρ_s is not considered fixed, but is an independent variable. Its rate of approach to its equilibrium value is determined by Eq. 17-25. The coefficient Λ determines the absorption of sound near the λ-point. It is simply related to the coefficients of second viscosity. According to experimental data, Λ is approximately equal to 15. The function $E_0(\rho, S, \rho_s)$ may, in principle, be determined from experimental data on the dependence of the specific heat and superfluid density ρ_s of helium on the temperature and pressure near the λ-point. The region of validity of the above equations is limited by the inequality $T_\lambda - T \ll T_\lambda$. The most interesting application of these equations is to processes in the presence of vortices, in which there are large gradients of ρ_s (cf. Ref. 22).

PART III

KINETIC PHENOMENA

18

THE KINETIC EQUATION FOR ELEMENTARY EXCITATIONS[13]

The distribution function of the elementary excitations satisfies the kinetic equation

$$\frac{\partial n}{\partial t} + \frac{\partial n}{\partial \mathbf{r}} \cdot \frac{\partial H}{\partial \mathbf{p}} - \frac{\partial n}{\partial \mathbf{p}} \cdot \frac{\partial H}{\partial \mathbf{r}} = \mathcal{I}(n) \tag{18-1}$$

where $\mathcal{I}(n)$ is the collision integral whose concrete form is not essential for us at this time. The Hamiltonian H, of a quasi-particle in the presence of superfluid motion with velocity $\mathbf{v}_s$ is equal to

$$H = \epsilon(p) + \mathbf{p} \cdot \mathbf{v}_s \tag{18-2}$$

where $\epsilon(p)$ is the energy in the frame of reference in which the superfluid component is at rest. Starting from the kinetic equation one may derive the hydrodynamic equations for a superfluid. In doing this one encounters expressions for the momentum and energy fluxes, which are just those needed for calculating the kinetic coefficients. Let us multiply Eq. 18-1 by the momentum component p_i and integrate over all momentum space. According to the law of conservation of energy, the integral on the right-hand side vanishes, so that we obtain

$$\frac{\partial}{\partial t} \int n p_i \, d\tau_p + \int p_i \frac{\partial n}{\partial r_k} \frac{\partial H}{\partial p_k} \, d\tau_p - \int p_i \frac{\partial n}{\partial p_k} \frac{\partial H}{\partial r_k} \, d\tau_p = 0$$

which after some simple manipulations yields

$$\frac{\partial}{\partial t}\overline{np_i} + \frac{\partial}{\partial r_k}\overline{np_i\frac{\partial H}{\partial p_k}} + \overline{n\frac{\partial H}{\partial r_i}} = 0 \tag{18-3}$$

Here and in what follows a bar signifies integration over momentum space. Inserting Eq. 18-2 into Eq. 18-3 we obtain the equation of motion

$$\frac{\partial}{\partial t}\overline{np_i} + \frac{\partial}{\partial r_k}\overline{np_i\left(\frac{\partial \epsilon}{\partial p_k} + v_{sk}\right)} + \overline{n\left(\frac{\partial \epsilon}{\partial r_i} + \frac{\partial}{\partial r_i}\mathbf{p}\cdot\mathbf{v}_s\right)} = 0 \tag{18-4}$$

which determines the time rate of change of the momentum of relative motion of the superfluid and normal components.

Let us now write the equation of motion for the total momentum

$$\mathbf{j} = \overline{\mathbf{p}n} + \rho\mathbf{v}_s \tag{18-5}$$

Since the total momentum is conserved its time derivative is equal to the divergence of a symmetric tensor, the momentum flux Π_{ik} (cf. Eq. 8-3)

$$\frac{\partial}{\partial t}j_i + \frac{\partial \Pi_{ik}}{\partial r_k} = 0 \tag{18-6}$$

As in Eq. 8-8 we may write the momentum flux tensor Π_{ik} in the rest frame in terms of its value π_{ik} in the frame of reference moving with the velocity $\mathbf{v}_s$

$$\Pi_{ik} = \pi_{ik} + v_{sk}\overline{np_i} + v_{si}\overline{np_k} + \rho v_{si}v_{sk} \tag{18-7}$$

Let us subtract Eq. 18-4 from Eq. 18-6 and use the continuity equation

$$\frac{\partial \rho}{\partial t} + \frac{\partial}{\partial r_k}(\overline{np_k} + \rho v_{sk}) \tag{18-8}$$

Finally we obtain

$$\rho\frac{\partial v_{si}}{\partial t} + \rho v_{sk}\frac{\partial v_{si}}{\partial r_k} + \frac{\partial \pi_{ik}}{\partial r_k} - \overline{n\frac{\partial \epsilon}{\partial r_i}} - \frac{\partial}{\partial r_k}\overline{np_i\frac{\partial \epsilon}{\partial p_k}} = 0 \tag{18-9}$$

From the condition curl $\mathbf{v}_s = 0$ it follows that the sum of the last three terms in Eq. 18-9 should be the product of the density ρ with the gradient of some function. The derivative $\partial\epsilon/\partial r_i$ may obviously be written in the form $\partial\epsilon/\partial\rho\ \partial\rho/\partial r_i$. Furthermore at absolute zero, in the absence of excitations, the tensor π_{ik} should be equal to $\rho_0\ \delta_{ik}$ (p_0 is the pressure in superfluid helium at T = 0). From these requirements, the form of the tensor π_{ik} follows uniquely

$$\pi_{ik} = \overline{np_i \frac{\partial\epsilon}{\partial p_k}} + \delta_{ik}\ p_0 + \overline{n\frac{\partial\epsilon}{\partial\rho}\rho} \tag{18-10}$$

According to the thermodynamic identity, Eq. 8-26, we have

$$\frac{\partial p_0}{\partial r_i} = \rho\ \frac{\partial\mu_0}{\partial r_i} \tag{18-11}$$

(μ_0 is the value of the chemical potential at T = 0).

If we insert Eq. 18-10 into Eq. 18-9 and take into account Eq. 18-11 we obtain

$$\frac{\partial\mathbf{v}_s}{\partial t} + \nabla\left\{\mu_0 + \overline{n\frac{\partial\epsilon}{\partial\rho}} + \frac{v_s^2}{2}\right\} = 0 \tag{18-12}$$

In this manner we have derived the equation of superfluid motion. Comparing Eqs. 18-12 and 8-25 we may see that the chemical potential μ is equal to

$$\mu = \mu_0 + \int n\frac{\partial\epsilon}{\partial\rho}\,d\tau_p \tag{18-13}$$

The second term in this formula is due to the excitations.

Let us now derive the expression for the energy flux vector **Q**. To do this we calculate the time derivative of the total energy and re-express all the time derivatives in terms of derivatives with respect to the coordinates, by means of the equations of motion, the equation of continuity and the kinetic equation. If we then group all terms into a divergence we find an expression for **Q**. The total energy consists of the kinetic energy (cf. Eq. 8-9)

$$E_k = \rho\frac{v_s^2}{2} + \mathbf{v}_s\cdot\overline{n\mathbf{p}} \tag{18-14}$$

the internal energy of the excitation gas,

$$E_k = \overline{n\epsilon} \tag{18-15}$$

and the zero-point energy E_0 (at T = 0), determined by the identity

$$dE_0 = \mu_0 \, d\rho \tag{18-16}$$

The time derivative of the kinetic energy is by Eqs. 18-4, 18-8, and 18-12 equal to

$$\begin{aligned}
\frac{\partial}{\partial t} E_k = & -(\rho v_{si} + \overline{np_i}) \frac{\partial}{\partial r_i}\left(\mu_0 + \overline{\frac{\partial \epsilon}{\partial \rho} n} + \frac{v_s^2}{2}\right) \\
& - \frac{v_s^2}{2}\frac{\partial}{\partial r_i}(\rho v_{si} + \overline{np_i}) - v_{si}\frac{\partial}{\partial r_k}\overline{np_i\left(\frac{\partial \epsilon}{\partial p_k} + v_{sk}\right)} \\
& - v_{si}\left(\overline{n\frac{\partial \epsilon}{\partial r_i}} + \overline{np_k}\frac{\partial v_{sk}}{\partial r_i}\right)
\end{aligned} \tag{18-17}$$

In order to find the time derivative of the internal energy we multiply both sides of the kinetic equation by ϵ and integrate over all space. The integral of the right-hand side vanishes since energy is conserved during collisions and we obtain

$$\overline{\epsilon \frac{\partial n}{\partial t}} + \overline{\epsilon \frac{\partial n}{\partial r_k}\frac{\partial H}{\partial p_k}} - \overline{\epsilon \frac{\partial n}{\partial p_k}\frac{\partial H}{\partial r_k}} = 0 \tag{18-18}$$

which may be transformed to

$$\begin{aligned}
& \frac{\partial}{\partial t}\overline{n\epsilon} + \frac{\partial}{\partial r_k}\overline{n\epsilon\left(\frac{\partial \epsilon}{\partial p_k} + v_{sk}\right)} - v_{sk}\overline{n\frac{\partial \epsilon}{\partial r_k}} - \overline{n\frac{\partial \epsilon}{\partial p_k}p_i}\frac{\partial v_{si}}{\partial r_k} \\
& \quad + \overline{n\frac{\partial \epsilon}{\partial \rho}}\frac{\partial}{\partial r_k}(\overline{np_k} + \rho v_{sk}) = 0
\end{aligned} \tag{18-19}$$

Making use of Eqs. 18-16, 18-17, and 18-19 we find the time derivative of the total energy

$$\frac{\partial E}{\partial t} = -\frac{\partial}{\partial r_k}\left\{(\overline{np_k} + \rho v_{sk})\left(\mu_0 + \overline{n\frac{\partial \epsilon}{\partial \rho}} + \frac{v_s^2}{2}\right) + \overline{nH\frac{\partial}{\partial p_k}H}\right\} \tag{18-20}$$

Thus the energy flux vector is equal to

$$\mathbf{Q} = (\overline{n\mathbf{p}} + \rho \mathbf{v}_s)\left(\mu_0 + \overline{n\frac{\partial \epsilon}{\partial \rho}} + \frac{v_s^2}{2}\right) + \overline{nH\frac{\partial}{\partial \mathbf{p}}H} \tag{18-21}$$

THE KINETIC COEFFICIENTS

The distribution function of the excitations in uniform motion and in an equilibrium state is equal to

$$n_0 = \left[\exp\left(\frac{\epsilon + \mathbf{p}\cdot\mathbf{v}_s - \mathbf{p}\cdot\mathbf{v}_n}{kT}\right) - 1\right]^{-1}$$

The equilibrium function n_0 satisfies the kinetic equation with vanishing collision integral. When the equilibrium is destroyed the distribution function differs from its equilibrium value and is found by solving the kinetic equation. This problem, which is in general not soluble is simplified when one considers states that differ only little from equilibrium. In this case the departure from equilibrium is completely determined by the first derivatives of the velocities $\mathbf{v}_n$ and $\mathbf{v}_s$ and the thermodynamic variables, with respect to the coordinates, all of which are assumed to be small quantities. In other words, the velocities $\mathbf{v}_n$ and $\mathbf{v}_s$ and the thermodynamic variables are slowly varying functions throughout the system, so that one may neglect all higher derivatives and higher powers of the first derivatives in the kinetic equation.

Let us consider a state of the system that differs little from the equilibrium state. We may then assume that in any small portion of the system we have an approximate local equilibrium—that is, that the distribution function may be represented in the form

$$n = n_0 + n_1 \tag{18-23}$$

where $n_1 \ll n_0$ and n_0 is the equilibrium function, Eq. 18-22, depending on the local values of the velocities and thermodynamic variables. When inserting the distribution function, Eq. 18-22, into the left-hand side of the kinetic equation, 18-21, it is sufficient to differentiate the function n_0, since the small additive contribution n_1 itself contains first derivatives and differentiating it would only bring in higher derivatives which may be neglected. The right-hand side, the collision integral, vanishes for $n = n_0$, and we may limit ourselves to terms linear in n_1. In this manner the problem has been linearized; we have obtained a linear integral equation whose left-hand side is known, and is expressible in terms of the first derivatives of the velocities and the thermodynamic variables. Let us insert the function n_0 into the left-hand side of the kinetic equation. Along with the spatial derivatives we shall consider that the velocity difference $\mathbf{v}_n - \mathbf{v}_s$ is small. This in no way limits the validity of our analysis since the velocity difference $\mathbf{v}_n - \mathbf{v}_s$ should in any case be small compared to the velocity of first and second sound. As is well known, superfluidity is destroyed long before this limit is reached. Let us first compute the time derivative; according to Eq. 18-1 we have†

$$\frac{\partial n_o}{\partial t} = n' \left\{ \frac{1}{kT} \left(\frac{\partial \epsilon}{\partial t} - \mathbf{p} \cdot \frac{\partial \mathbf{v}_n}{\partial t} + \mathbf{p} \cdot \frac{\partial \mathbf{v}_s}{\partial t} \right) - \frac{\epsilon}{kT^2} \frac{\partial T}{\partial t} \right\} \qquad (18\text{-}24)$$

Let us choose as our independent variables the density ρ and the entropy S. We may then express the time derivatives in terms of space derivatives; in the linear approximation we have, according to Eqs. 8-22 to 8-25,

$$\frac{\partial \rho}{\partial t} = -\operatorname{div} \mathbf{j} \qquad \frac{\partial S}{\partial t} = -S \operatorname{div} \mathbf{v}_n \qquad \rho_n \frac{\partial}{\partial t} (\mathbf{v}_n - \mathbf{v}_s) = -S\nabla T \qquad (18\text{-}25)$$

By using Eq. 18-25 we may transform Eq. 18-24 to

$$\frac{\partial n_o}{\partial t} = \frac{n'}{kT} \left\{ \operatorname{div} (\mathbf{j} - \rho \mathbf{v}_n) \left(\frac{1}{T} \frac{\partial T}{\partial \rho} \epsilon - \frac{\partial \epsilon}{\partial \rho} \right) + \operatorname{div} \mathbf{v}_n \times \left[\frac{1}{T} \left(\frac{\partial T}{\partial \rho} \rho + \frac{\partial T}{\partial S} S \right) \epsilon - \frac{\partial \epsilon}{\partial \rho} \rho \right] + \frac{ST}{\rho_n} \mathbf{p} \cdot \frac{\nabla T}{T} \right\} \qquad (18\text{-}26)$$

Let us now calculate the Poisson bracket on the left-hand side of the kinetic equation

$$\frac{\partial n}{\partial \mathbf{r}} \cdot \frac{\partial H}{\partial \mathbf{p}} - \frac{\partial n}{\partial \mathbf{p}} \cdot \frac{\partial H}{\partial \mathbf{r}} = \frac{n'}{kT} \left\{ -\frac{\partial \epsilon}{\partial \mathbf{p}} \cdot \nabla (\mathbf{p} \cdot \mathbf{v}_n) - \frac{\nabla T}{T} \epsilon \cdot \frac{\partial \epsilon}{\partial \mathbf{p}} \right\} \qquad (18\text{-}27)$$

Collecting all the terms in Eq. 18-26 and 18-27 we obtain the kinetic equation in the approximation which interests us

$$\frac{n'}{kT} \left\{ \left(\frac{1}{T} \frac{\partial T}{\partial \rho} \epsilon - \frac{\partial \epsilon}{\partial \rho} \right) \operatorname{div} (\mathbf{j} - \rho \mathbf{v}_n) + \left[\frac{1}{T} \left(\frac{\partial T}{\partial \rho} \rho + \frac{\partial T}{\partial s} s \right) \epsilon - \frac{\partial \epsilon}{\partial \rho} \right] \operatorname{div} \mathbf{v}_n + \frac{\nabla T}{T} \cdot \left(\mathbf{p} \frac{ST}{\rho_n} - \epsilon \frac{\partial \epsilon}{\partial \mathbf{p}} \right) - \frac{\partial \epsilon}{\partial \mathbf{p}} \cdot \nabla (\mathbf{p} \cdot \mathbf{v}_n) \right\} = \mathcal{I}(n_1) \qquad (18\text{-}28)$$

---

†We denote with a prime differentiation of a function with respect to its argument, $n' = -n(1 + n)$.

For what follows it is convenient to separate from the $\partial\epsilon/\partial\mathbf{p}\cdot\nabla(\mathbf{p}\cdot\mathbf{v}_n)$ term in the curly brackets a term of the form div $\mathbf{v}_n$, and to symmetrize the rest; this yields

$$\frac{n'}{kT}\left\{\left(\frac{1}{T}\frac{\partial T}{\partial\rho}\epsilon-\frac{\partial\epsilon}{\partial\rho}\right)\operatorname{div}(\mathbf{j}-\rho\mathbf{v}_n)+\left[\frac{1}{T}\left(\frac{\partial T}{\partial\rho}\rho+\frac{\partial T}{\partial S}S\right)\epsilon\right.\right.$$

$$\left.-\frac{\partial\epsilon}{\partial\rho}-\frac{1}{3}\frac{\partial\epsilon}{\partial p}p\right]\operatorname{div}\mathbf{v}_n+\frac{\nabla T}{T}\cdot\left(\mathbf{p}\frac{ST}{\rho_n}-\epsilon\frac{\partial\epsilon}{\partial\mathbf{p}}\right)$$

$$-\left(\frac{\partial\epsilon}{\partial p_i}p_k-\frac{1}{3}\delta_{ik}\frac{\partial\epsilon}{\partial p}p\right)\left(\frac{\partial v_{ni}}{\partial r_k}+\frac{\partial v_{nk}}{\partial r_i}\right.$$

$$\left.\left.-\frac{2}{3}\delta_{ik}\operatorname{div}\mathbf{v}_n\right)\right\}=\mathscr{I}(n_1) \qquad (18\text{-}29)$$

An analysis of Eq. 18-29 shows that, in agreement with Chap. 9, terms containing div$(\mathbf{j}-\rho\mathbf{v}_n)$ and div $\mathbf{v}_n$ determine the second viscosity in the superfluid (in all, this involves three coefficients); the term with a temperature gradient ∇T determines the heat conduction, and finally the last term determines the first viscosity. We shall perform the calculation of the corresponding kinetic coefficients below, using the general equation, 18-29.

19

THERMAL CONDUCTIVITY

When a temperature gradient is present in superfluid helium, there arises in the liquid a macroscopic motion which may be described by using the equations of motion. The macroscopic motion of the normal fluid is accompanied by a transport of heat in the opposite direction. Moreover, apart from this macroscopic heat flux there is also an irreversible heat flow, which according to the analysis of Chap. 9 is expressible in terms of the coefficient of thermal conduction κ by the formula

$$\mathbf{q} = -\kappa \nabla T \tag{19-1}$$

In order to calculate the coefficient κ it is necessary to solve the kinetic equation, which in the presence of a temperature gradient may be Eq. 18-29 be written in the form

$$\frac{n'}{\kappa T} \nabla T \cdot \left(\mathbf{p} \frac{ST}{\rho_n} - \epsilon \frac{\partial \epsilon}{\partial \mathbf{p}} \right) = \mathcal{J}(n_1) \tag{19-2}$$

The phenomenon of thermal conduction we are here considering has much in common with thermal transport in a classical liquid. However, in this case there are certain specific features connected with the fact that the effect is due to excitations having an unusual energy spectrum. Indeed, for a pure phonon gas, the left-hand side of Eq. 19-2 is identically zero and consequently the corresponding coefficient of thermal conduction would vanish.

We may perform our calculation by separating the current Eq. 19-1 into two parts, due to rotons and phonons, respectively. The coefficient of thermal conducttion κ consists also of two parts, the roton part κ_r and the phonon part κ_{ph}

$$\kappa = \kappa_r + \kappa_{ph}$$

THE ROTON PART OF THE THERMAL CONDUCTIVITY

The coefficient κ_r is determined primarily by roton-roton collision processes. For relatively high temperatures, (above 0.9°K), the interaction of phonons with rotons turns out to be negligible. For lower temperatures this interaction begins to play a role; however, as we shall see, in this region of temperatures the main contribution to the thermal conduction comes from the phonons, so that we may merely neglect this roton-phonon effect. We shall only calculate κ_r in order of magnitude in order to determine the temperature dependence of this coefficient. There is no point in attempting a more accurate calculation since the character of the roton-roton interaction is not known (Chap. 7). For our purposes it is possible to simplify the problem by replacing the collision integral by the quantity

$$\mathcal{J}(n) \to \frac{n - n_0}{t_r} \tag{19-3}$$

where t_r is some characteristic time which we shall later identify with the collision time of rotons determined by formula 7-40. Let us insert Eq. 19-3 into the right-hand side of Eq. 19-2 and solve the kinetic equation for $n - n_0$; this yields

$$n - n_0 = -\frac{n'}{kT^2} \nabla T \cdot \left(p \frac{ST}{\rho_n} - \epsilon \frac{\partial \epsilon}{\partial p}\right) t_r \tag{19-4}$$

Let us futher calculate the energy flux

$$\mathbf{q} = \int \frac{\partial \epsilon}{\partial \mathbf{p}} \epsilon (n - n_0)\, d\tau_p \tag{19-5}$$

By comparing Eq. 19-5 and 19-1 we find an expression for the coefficient of thermal conduction

$$\kappa_r = t_r \frac{1}{3kT^2} \int n' \epsilon \frac{\partial \epsilon}{\partial p} \cdot \left(p \frac{ST}{\partial_n} - \epsilon \frac{\partial \epsilon}{\partial p}\right) d\tau_p \tag{19-6}$$

If we calculate the integral in Eq. 19-6 with the roton distribution function given in Eq. 1-14 we obtain the following result

$$\kappa_r = \frac{\Delta^2 t_r \mathfrak{N}_r}{3\mu T}\left(1 - \frac{3SS_r T^2 \mu}{\rho_n N_r}\right) \tag{19-7}$$

The second term in the bracket in Eq. 19-7 is in practice always negligible. Inserting expression 7-40 for t_r into Eq. 19-7 we finally have

$$\kappa_r = \frac{\Delta^2 \hbar^4_r}{12 p_0 \mu^2 |V_0|^2} \cdot \frac{1}{T} \tag{19-8}$$

Thus the roton part of the coefficient of thermal conduction increases slowly with decreasing temperature, as the inverse power of T.

THE PHONON PART OF THE THERMAL CONDUCTIVITY

A study of the phonon contribution to the thermal conductivity reveals that the main contribution comes from phonon-roton scattering processes.

The solution of the kinetic equation is greatly simplified by the fact that energy equilibrium is established in the phonon gas in times which are much shorter than the pnonon-roton scattering times for the processes mentioned above. The establishment of energy equilibrium is due to phonon-phonon scattering at small angles, for which the scattering times were calculated in Chap. 7. These processes do not contribute to transport phenomena, such as thermal conduction or viscosity; however they do bring phonons having a given direction of momentum, into energy equilibrium. This allows us to describe the phonons by means of equilibrium distribution functions, with each momentum direction characterized by a definite temperature.

Furthermore, it must be noted that equilibrium in the number of phonons is established by means of a five-phonon process. This process (2 phonons $\rightleftharpoons$ 3 phonons), as we saw previously, also takes place without change in the direction of the colliding phonons. For $T > 0.9°K$, the times characterizing this process are comparable with the phonon-roton scattering time. For $T < 0.9°K$ the rotons rapidly cease to play a role, and the establishment of equilibrium in the number of phonons takes place more rapidly than the scattering of phonons. Thus the phonon distribution function for a given direction contains a chemical potential which is a function of direction (below 0.9°K it may be set identically to zero). The phonons moving in a given direction have a distribution function which may be written in the form

$$n = \left[\exp\left(\alpha' + \frac{pc}{kT'}\right) - 1\right]^{-1} \tag{19-9}$$

where α' is some function of direction which is analogous to a chemical potential and T' is a temperature. The distribution function given in Eq. 19-9 differs from its equilibrium value n_0 by the quantity

$$n - n_0 = -n_0(n_0 + 1)\left\{\alpha' - \frac{pc}{kT}\frac{T' - T}{T}\right\} \tag{19-10}$$

T is the equilibrium temperature of the phonon gas). Let us choose as our z-axis the direction of ∇T. Then the left-hand side of the kinetic equation contains a factor $\cos\theta$, where θ is the angle between the phonon momentum and the z-axis. Since Eq. 19-2 is linear the angular dependence of the functions α' and $(T' - T)/T$ is obviously

$$\alpha' = \alpha\cos\theta \qquad \frac{T' - T}{T} = \beta\cos\theta \tag{19-11}$$

The collision integral on the right-hand side of Eq. 19-2 is composed of four terms, corresponding to the different processes that can occur: $\mathcal{J}_{r\,ph}$ for phonon-roton collisions, $\mathcal{J}_{pp}$ for phonon-phonon collision, $\mathcal{J}_{\mathcal{g}\mathcal{V}}$ for small-angle phonon-phonon collisions, and $\mathcal{J}_{\mathcal{V}}$ for the collision of two phonons accompanied by the creation of a third phonon (the five-phonon process).† Let us integrate Eq. 19-2 over all phonons and overall phonon energies $\int \mathcal{J}p^2\,dp$ and $\int \mathcal{J}\epsilon p^2\,dp$). The foregoing discussion shows us that the integrals $\int \mathcal{J}_{\mathcal{g}\mathcal{V}}p^2\,dp$, $\int \mathcal{J}_{\mathcal{g}\mathcal{V}}\epsilon p^2\,dp$, and $\int \mathcal{J}_{\mathcal{V}}\epsilon p^2\,dp$ are zero.

The integral of $\mathcal{J}_{pp}$ may also be shown to vanish identically. This is obvious from physical considerations, since in phonon-phonon collisions momentum is conserved so that there can be no flux.

According to Eq. 7-44 we have

$$\frac{1}{kT}\int \mathcal{J}_{\mathcal{V}}\,d\tau_p = \alpha\Gamma_{ph} \tag{19-12}$$

The integral of $\mathcal{J}_{r\,ph}$ may be calculated simply by using formula 7-31 for the differential scattering cross section of phonons by rotons

$$\mathcal{J}_{r\,ph} = \cos\theta\,\mathfrak{N}_r c\int d\sigma(p,\psi)(1 - \cos\psi)\left(\alpha - \beta\frac{pc}{kT}\right)n_0(n_0 + 1)$$

---

†See footnote on page 49.

We may integrate this expression over the solid angle do to obtain

$$\mathscr{g}_{\mathrm{r\,ph}} = n_0(n_0 + 1)\left(\alpha - \beta\frac{pc}{kT}\right)\cos\vartheta\,\frac{1}{6!\theta}\left(\frac{pc}{kT}\right)^4 \qquad (19\text{-}13)$$

where θ has the dimensions of time and characterize the scattering of phonons by rotons

$$\frac{1}{\theta} = \frac{6!\mathfrak{N}_r}{4\pi c}\left[\frac{p_0(kT/c)^2}{\hbar^2\rho}\right]^4\left\{\frac{4}{45} + \frac{1}{25}\left(\frac{p_0}{\mu c}\right)^2 + \frac{4}{9}\frac{p_0}{\mu c}A + A^2\right\} \qquad (19\text{-}14)$$

Let us use the formulas given in Eqs. 19-10 to 19-13, after having taken averages over all phonons and over all phonon energies (for a given direction θ). We may insert these into the kinetic equation to obtain two equations determining the quantities α and β

$$\frac{1}{T}\frac{\partial T}{\partial x}\frac{4\pi^4}{15}\left(c - \frac{ST}{\rho_n c}\right) = \frac{7}{\theta}(\alpha - 8\beta)$$

$$\frac{1}{T}\frac{\partial T}{\partial x}\frac{36}{5}\left(c - \frac{ST}{\rho_n c}\right) = \frac{1}{\theta}(\alpha - 7\beta) + \frac{1}{\theta_{ph}}\alpha \qquad (19\text{-}15)$$

here θ_{ph} is the time characterizing the five-phonon process

$$\frac{1}{\theta_{ph}} = \frac{(2\pi\hbar)^3}{4\pi(kT/c)^3}kT\,\Gamma_{ph} \qquad (19\text{-}16)$$

Having obtained the quantities α and β we know, by Eq. 19-10, the nonequilibrium part of the distribution function, and we may easily calculate the energy flux and consequently the phonon contribution to the thermal conductivity.

It must be noted, however, that the solution obtained for the kinetic equation, 19-10, does not satisfy the condition $\int \mathbf{p}(n - n_0)\,d\tau = 0$. Therefore in the presence of a temperature gradient there occurs a macroscopic motion of the liquid with relative velocity $\mathbf{v}_n - \mathbf{v}_s$, such that

$$\int \mathbf{p}(n - n_0)\,d\tau + \rho_n(\mathbf{v}_n - \mathbf{v}_s) = 0$$

Thus the equilibrium distribution functions of phonons and rotons will not only depend on the energy ϵ, but on the combination $\epsilon - \mathbf{p}\cdot(\mathbf{v}_n - \mathbf{v}_s)$. Expanding the equilibrium functions in a series in $\mathbf{v}_n - \mathbf{v}_s$ we obtain an additional term $-(\partial n/\partial\epsilon)\mathbf{p}\cdot(\mathbf{v}_n - \mathbf{v}_s)$, where

$v_n - v_s$ is determined by the previous condition. The total addition to the phonon distribution function is thus equal to

$$\delta n = (n - n_0) + \frac{\partial n}{\partial \epsilon} \frac{1}{\rho_n} \mathbf{p} \cdot \int (n' - n_0) \mathbf{p}' \, d\tau'$$

where the difference $n - n_0$ is determined by formula 19-10.

In a similar manner we obtain the addition to the roton distribution function

$$\delta \mathfrak{N}_r = \frac{\partial \mathfrak{N}_r}{\partial E} \frac{1}{\rho_n} \mathbf{p} \cdot \int (n - n_0) \mathbf{p}' \, d\tau'$$

This leads to the following expression for the energy current caused by phonons (and rotons)

$$\mathbf{Q} = \int \delta n \epsilon \frac{\partial \epsilon}{\partial \mathbf{p}} \, d\tau + \int \delta \mathfrak{N}_r E \frac{\partial E}{\partial \mathbf{p}} \, d\tau$$

$$= \left(1 - \frac{ST}{\rho_n c^2}\right) \int (n - n_0) \epsilon \frac{\partial \epsilon}{\partial \mathbf{p}} \, d\tau$$

After a simple calculation we obtain the final result

$$\kappa_{ph} \approx 2\theta \, \mathfrak{N}_{ph} \kappa \left(1 - \frac{ST}{\rho_n c^2}\right)^2 \left(\frac{10 + (8\theta/\theta_{ph})}{1 + (8\theta/\theta_{ph})}\right) \tag{19-17}$$

Since the number of rotons $\mathfrak{N}_r$ depends exponentially on the temperature, the time θ increases rapidly with decreasing temperature, so that the main temperature dependence of κ_{ph} is as $e^{\Delta/kT}$. Below 1.4°K the phonon part of the thermal conductivity κ_{ph} is already considerably greater than the roton part (κ_r). For temperatures below 0.9°K, as we mentioned before, we have $\theta \gg \theta_{ph}$ and the last parenthesis in Eq. 19-17 can be replaced by unity.

20

THE FIRST VISCOSITY

The problem of calculating the coefficient of first viscosity is very similar to the one we just discussed concerning thermal conductivity. Let us consider the properties of superfluid helium, in which the normal velocity $\mathbf{v}_n$ is a function of the coordinates.

For the calculation of the viscosity the kinetic equation, 18-29 has the following form:

$$-\frac{n'}{kT}\left(\frac{\partial\epsilon}{\partial p_i}\,p_k - \tfrac{1}{3}\,\delta_{ik}\,\frac{\partial\epsilon}{\partial \mathbf{p}}\cdot\mathbf{p}\right)\left(\frac{\partial v_{ni}}{\partial r_k} + \frac{\partial v_{nk}}{\partial r_i} - \tfrac{2}{3}\,\delta_{ik}\,\frac{\partial v_{n\ell}}{\partial r_\ell}\right) = \mathcal{I}(n) \tag{20-1}$$

We must calculate the momentum flux which arises, and for this purpose we divide it into two parts, one referrring to rotons and the other to phonons.

As in Chap. 19 we may calculate the roton part of the viscosity coefficient η_r by changing $\mathcal{I}(n)$ into

$$-\frac{n - n_0}{t_r} \tag{20-2}$$

The time t_r which occurs in this formula is in general not the same as the time t_r used in Chap. 19 for the calculation of the roton part of the thermal conductivity. The two times would only coincide if the roton-roton scattering amplitude had a narrow maximum for small angles. However, since we are only interested in the temperature dependence of η_r, the approximation made in Eq. 20-2, with t_r given by 7-40, will be suffient. Let us insert Eq. 20-2 into the right-hand side of Eq. 20-1, calculate $n - n_0$, and then find the momentum flux tensor; this yields

$$\pi_{\ell m} = \int \frac{\partial \epsilon}{\partial p_\ell} p_m (n - n_0)\, d\tau_p = -\frac{t_r}{15kT} \int n' \left(\frac{\partial \epsilon}{\partial p}\right)^2 p^2$$

$$\times\, d\tau_p \cdot \left(\frac{\partial v_{n\ell}}{\partial r_m} + \frac{\partial v_{nm}}{\partial r_\ell} - \tfrac{2}{3}\,\delta_{\ell m}\, \mathrm{div}\, \mathbf{v}_n\right) \qquad (20\text{-}3)$$

Performing the necessary integration in Eq. 20-3 we find for the roton part of the viscosity

$$\eta_r = \frac{p_0^2 t_r \mathfrak{N}_r}{15\mu} = \frac{\hbar^2 p_0}{60\mu^2 |V_0|^2} \qquad (20\text{-}4)$$

Expression 20-4 contains no terms depending on the temperature, so that η_r turns out to be a constant.

The phonon part of the viscosity η_{ph} may be calculated in the same way as the phonon part of the thermal conductivity. There are only two points of difference. First, because of the different symmetry of the problem, the functions α' and $(T' - T)/T$ (see formula 19-10) will have the form

$$\alpha' = \alpha P_2(\cos\theta) \qquad \frac{T' - T}{T} = \beta P_2(\cos\theta) \qquad (20\text{-}5)$$

Second, phonon-phonon scattering at large angles does give a contribution for temperatures below 0.9°K (the integral of $\mathcal{J}_{pp}$ does not vanish).

We shall not perform the detailed calculations here but only write down the final result

$$\eta_{ph} \approx 0.4\,\mathfrak{N}_{ph}\, kT\bar{\theta} \begin{cases} \dfrac{10 + (8\bar{\theta}/\theta_{ph})}{1 + (8\bar{\theta}/\theta_{ph})} \\[2ex] \dfrac{1}{1 + (\bar{\theta}/56\tau_{ph})} \end{cases} \qquad (20\text{-}6)$$

Here the time θ_{ph} is determined by Eq. 19-16, whereas the time $\overline{\theta}$ is slightly different from θ (formula 19-14)

$$\frac{1}{\overline{\theta}} = \frac{\mathfrak{N}_r\, 6!}{4\pi c}\left[\frac{p_0 (kt/c)^2}{\hbar^2 \rho}\right]^2 \left\{\frac{2}{15} + \frac{33}{(35)^2}\left(\frac{p_0}{\mu c}\right)^2 + \frac{4A}{75}\left(\frac{p_0}{\mu c}\right) + A^2\right\} \tag{20-7}$$

and finally the time τ_{ph} characterizing the four-phonon process is equal to

$$\frac{1}{\tau_{ph}} = \frac{3.13!(u+2)^4}{5.2^{19}(2\pi)^3\hbar^7\rho^2 c}\left(\frac{kT}{c}\right)^9 \qquad \left(u = \frac{2\rho}{c}\frac{\partial c}{\partial \rho}\right) \tag{20-8}$$

A numerical analysis of Eq. 20-6 gives the following result: the phonon part of the viscosity increases exponentially with decreasing temperature as $\epsilon^{\Delta/kT}$; for temperatures below 0.7°K, when the only important effect comes from phonon-phonon scattering, this temperature dependence is changed to T^{-5}. For temperatures below 1.4°K the constant contribution of the roton viscosity is negligibly small in comparison with the phonon part.

21

THE COEFFICIENTS OF SECOND VISCOSITY OF HELIUM II

When the second viscosity is taken into account, the equations of motion of helium II are, according to Eqs. 9-14 and 9-15

$$\frac{\partial \mathbf{j}}{\partial t} + \nabla p + \mathbf{v_s}\,\mathrm{div}\,\mathbf{j} + (\mathbf{j} - \rho\mathbf{v_s})\,\mathrm{div}\,\mathbf{v_n} + (\mathbf{j}\cdot\nabla)\mathbf{v_s}$$

$$+ (\mathbf{v_n}\cdot\nabla)(\mathbf{j} - \rho\mathbf{v_n}) = \nabla\{\zeta_1\,\mathrm{div}(\mathbf{j} - \rho\mathbf{v_n}) + \zeta_2\,\mathrm{div}\,\mathbf{v_n}\} \qquad (21\text{-}1)$$

$$\frac{\partial \mathbf{v_s}}{\partial t} + \left(\mu + \frac{\mathbf{v_s^2}}{2}\right) = \nabla\{\zeta_3\,\mathrm{div}(\mathbf{j} - \rho\mathbf{v_n}) + \zeta_4\,\mathrm{div}\,\mathbf{v_n}\} \qquad (21\text{-}2)$$

The unusual character of the hydrodynamics of helium II leads to the appearance of four second viscosity coefficients. According to the Onsager reciprocity principle, the coefficients ζ_1 and ζ_4 are equal. Let us find the dependence of the second viscosity coefficients on the temperature and the other thermodynamic quantities. In order to do this, we start from the fact that the second viscosity in helium II is due to processes in which the total number of phonons $\mathfrak{N}_{ph}$ and of rotons $\mathfrak{N}_r$ changes.

Let $\mathfrak{N}_r$ and $\mathfrak{N}_{ph}$ be the number of rotons and phonons, respectively, per unit volume, and μ_r and μ_{ph} their chemical potentials. In thermodynamic equilibrium when μ_r and μ_{ph} are equal to zero, the numbers of rotons and phonons are functions of the density ρ and the entropy S and are equal to $\mathfrak{N}_{r0}$ and $\mathfrak{N}_{ph\,0}$, respectively. In the system disturbed from equilibrium, the numbers $\mathfrak{N}_r$ and $\mathfrak{N}_{ph}$ will vary in time and will tend toward their equilibrium values $\mathfrak{N}_{r0}$ and $\mathfrak{N}_{ph\,0}$. Let us consider small deviations from equilibrium, for which the density and the entropy differ little from their constant equilibrium values. We may, without limiting the generality of our discussion, also consider the velocities $\mathbf{v}_n$ and $\mathbf{v}_s$ to be small. The equation characterizing the approach of the system to equilibrium may be obtained by expanding the rates of change of the roton and phonon numbers, $\dot{\mathfrak{N}}_r$ and $\dot{\mathfrak{N}}_{ph}$, in powers of the chemical potentials. If we limit ourselves to terms linear in μ_r and μ_{ph} the equations take the form

$$\dot{\mathfrak{N}}_r + \operatorname{div} \mathfrak{N}_r \mathbf{v}_n = -\gamma_{rr}\mu_r + \gamma_{r\,ph}\mu_{ph} \tag{21-3}$$

$$\dot{\mathfrak{N}}_{ph} + \operatorname{div} \mathfrak{N}_{ph} \mathbf{v}_n = \gamma_{ph\,r}\mu_r - \gamma_{pp}\mu_{ph} \tag{21-4}$$

where γ_{rr}, $\gamma_{r\,ph}$, $\gamma_{ph\,r}$, and γ_{pp} are kinetic coefficients which are symmetric in the indices r and ph. The terms of the form $\mathfrak{N}\mathbf{v}_n$ in Eqs. 21-3 and 21-4 reflect the fact that phonons and rotons participate in the normal motion, with velocity $\mathbf{v}_n$. Neglecting quadratic effects in Eqs. 21-3 and 21-4 we obtain

$$\dot{\mathfrak{N}}_r + \mathfrak{N}_r \operatorname{div} \mathbf{v}_n = -\gamma_{rr}\mu_r + \gamma_{rph}\mu_{ph} \tag{21-5}$$

$$\dot{\mathfrak{N}}_{ph} + \mathfrak{N}_{ph} \operatorname{div} \mathbf{v}_n = \gamma_{r\,ph}\mu_r - \gamma_{pp}\mu_{ph} \tag{21-6}$$

The changes in time of the thermodynamic quantities ρ and S are determined by the continuity equations for density and entropy, which in this case can be linearized similarly to Eqs. 21-5 and 21-6 to yield

$$\dot{\rho} + \operatorname{div} \mathbf{j} = 0$$
$$\dot{S} + \operatorname{div} \mathbf{v}_n = 0 \tag{21-7}$$

The numbers of rotons $\mathfrak{N}_r$ and of phonons $\mathfrak{N}_{ph}$ depend on three variables—ρ, S, μ_r or μ_{ph}. For sufficiently slow departures from equilibrium, the functions $\mathfrak{N}_r$ and $\mathfrak{N}_{ph}$ are able to follow the changes in the thermodynamic quantities. In that case the derivatives

$\dot{\mathfrak{N}}_r$ and $\dot{\mathfrak{N}}_{ph}$ are expressible in terms of the derivatives $\dot{\rho}$ and $\dot{S}$. Taking into account (21.7) we can rewrite (21.5) and (21.6) as

$$-\frac{\partial \mathfrak{N}_r}{\partial \rho}\operatorname{div}(\mathbf{j}-\rho\mathbf{v}_n)+\left(\mathfrak{N}_r-\frac{\partial \mathfrak{N}_r}{\partial S}S-\frac{\partial \mathfrak{N}_r}{\partial \rho}\rho\right)\operatorname{div}\mathbf{v}_n$$

$$=-\gamma_{rr}\mu_r+\gamma_{r\,ph}\mu_{ph} \tag{21-8}$$

$$-\frac{\partial \mathfrak{N}_{ph}}{\partial \rho}\operatorname{div}(\mathbf{j}-\rho\mathbf{v}_n)+\left(\mathfrak{N}_{ph}-\frac{\partial \mathfrak{N}_{ph}}{\partial S}S-\frac{\partial \mathfrak{N}_{ph}}{\partial \rho}\rho\right)\operatorname{div}\mathbf{v}_n$$

$$=\gamma_{r\,ph}\mu_r-\gamma_{pp}\mu_{ph} \tag{21-9}$$

These equations determine the chemical potentials μ_r and μ_{ph} in terms of $\operatorname{div}(\mathbf{j}-\rho\mathbf{v}_n)$ and $\operatorname{div}\mathbf{v}_n$ for sufficiently slow nonequilibrium processes.

The kinetic coefficients γ which occur in these equation can be simply expressed in terms of the coefficients Γ calculated in Chap. 7 for certain concrete processes. By comparing Eqs. 21-5 and 21-6 with Eqs. 7-42 and 7-40 and taking into account the definition of Γ, we find

$$\gamma_{rr}=\Gamma_r+\Gamma_{r\,ph}\qquad \gamma_{r\,ph}=\gamma_{ph\,r}=\Gamma_{r\,ph}\qquad \gamma_{pp}=\Gamma_{ph}+\Gamma_{r\,ph} \tag{21-10}$$

The coefficient Γ_r is small compared to $\Gamma_{r\,ph}$ and Γ_{ph}, and may therefore be neglected. Relations 21-8 and 21-9 then yield†

$$\mu_r=\frac{1}{\Gamma_{ph}}\left\{\frac{\partial \mathfrak{N}}{\partial \rho}\operatorname{div}(\mathbf{j}-\rho\mathbf{v}_n)-\left(\mathfrak{N}-\frac{\partial \mathfrak{N}}{\partial S}S-\frac{\partial \mathfrak{N}}{\partial \rho}\rho\right)\operatorname{div}\mathbf{v}_n\right\}$$

$$+\frac{1}{\Gamma_{r\,ph}}\left\{\frac{\partial \mathfrak{N}_r}{\partial \rho}\operatorname{div}(\mathbf{j}-\rho\mathbf{v}_n)\right.$$

$$\left.-\left(\mathfrak{N}_r-\frac{\partial \mathfrak{N}_r}{\partial S}S-\frac{\partial \mathfrak{N}_r}{\partial \rho}\rho\right)\operatorname{div}\mathbf{v}_n\right\} \tag{21-11}$$

$$\mu_{ph}=\frac{1}{\Gamma_{ph}}\left\{\frac{\partial \mathfrak{N}}{\partial \rho}\operatorname{div}(\mathbf{j}-\rho\mathbf{v}_n)-\left(\mathfrak{N}-\frac{\partial \mathfrak{N}}{\partial S}S-\frac{\partial \mathfrak{N}}{\partial \rho}\rho\right)\operatorname{div}\mathbf{v}_n\right\} \tag{21-12}$$

†Note that $N=\mathfrak{N}_r+\mathfrak{N}_{ph}$

The supplementary terms in the equations of motion, Eqs. 21-1 and 21-2 are due to processes in which the total numbers of rotons $\mathfrak{N}_r$ and phonons $\mathfrak{N}_{ph}$ change. Therefore these terms are connected with the dependence of the pressure p and the chemical potential μ on the potentials μ_{ph} and μ_r and are, therefore, equal to

$$\nabla\left(\frac{\partial p}{\partial \mu_r}\mu_r + \frac{\partial p}{\partial \mu_{ph}}\mu_{ph}\right) \qquad \nabla\left(\frac{\partial \mu}{\partial \mu_r}\mu_r + \frac{\partial \mu}{\partial \mu_{ph}}\mu_{ph}\right)$$

Inserting these expression, as well as those found for μ_r and μ_{ph} in Eqs. 21-11 and 21-12, into Eqs. 21-1 and 21-2 we obtain two equations. Since these equations should hold for arbitrary values of div $\mathbf{v}_n$ and div$(\mathbf{j} - \rho\mathbf{v}_n)$, we can write down the following expressions

$$\zeta_1 = -\frac{1}{\Gamma_{ph}}\left(\frac{\partial p}{\partial \mu_r} + \frac{\partial p}{\partial \mu_{ph}}\right)\frac{\partial \mathfrak{N}}{\partial \rho} - \frac{1}{\Gamma_{r\,ph}}\frac{\partial p}{\partial \mu_r}\frac{\partial \mathfrak{N}_r}{\partial \rho} \tag{21-13}$$

$$\zeta_2 = \frac{1}{\Gamma_{ph}}\left(\frac{\partial p}{\partial \mu_r} + \frac{\partial p}{\partial \mu_{ph}}\right)\left(\mathfrak{N} - \frac{\partial \mathfrak{N}}{\partial S}S - \frac{\partial \mathfrak{N}}{\partial \rho}\rho\right)$$
$$+ \frac{1}{\Gamma_{r\,ph}}\frac{\partial p}{\partial \mu_r}\cdot\left(\mathfrak{N}_r - \frac{\partial \mathfrak{N}_r}{\partial S}S - \frac{\partial \mathfrak{N}_r}{\partial \rho}\rho\right) \tag{21-14}$$

$$\zeta_3 = -\frac{1}{\Gamma_{ph}}\frac{\partial \mu}{\partial \mu_r} + \frac{\partial \mu}{\partial \mu_{ph}}\frac{\partial \mathfrak{N}}{\partial \rho} - \frac{1}{\Gamma_{r\,ph}}\frac{\partial \mu}{\partial \mu_r}\frac{\partial \mathfrak{N}_r}{\partial \rho} \tag{21-15}$$

$$\zeta_4 = \frac{1}{\Gamma_{ph}}\left(\frac{\partial \mu}{\partial \mu_r} + \frac{\partial \mu}{\partial \mu_{ph}}\right)\left(\mathfrak{N} - \frac{\partial \mathfrak{N}}{\partial S}S - \frac{\partial \mathfrak{N}}{\partial \rho}\rho\right)$$
$$+ \frac{1}{\Gamma_{r\,ph}}\frac{\partial \mu}{\partial \mu_r}\cdot\left(\mathfrak{N}_r - \frac{\partial \mathfrak{N}_r}{\partial S}S - \frac{\partial \mathfrak{N}_r}{\partial \rho}\rho\right) \tag{21-16}$$

The function E_0 may be determined by means of the following identity

$$dE_0 = T\,dS + \mu\,d\rho - \mathfrak{N}_r\,d\mu_r - \mathfrak{N}_{ph}\,d\mu_{ph} \tag{21-17}$$

Thus E_0 is analogous to an energy, and for constant chemical potentials it becomes the energy per unit volume of the liquid. The pressure p is expressible in terms of the energy E_0 by means of the relation

$$p = -E_0 + ST + \mu\rho \tag{21-18}$$

From Eqs. 21-17 and 21-18 we find the following expressions for the derivatives $\partial p/\partial\mu_r$ and $\partial p/\partial\mu_{ph}$

$$\frac{\partial p}{\partial\mu_r} = \mathfrak{N}_r - S\,\frac{\partial\mathfrak{N}_r}{\partial S} - \rho\,\frac{\partial\mathfrak{N}_r}{\partial\rho}$$

$$\frac{\partial p}{\partial\mu_{ph}} = \mathfrak{N}_{ph} - S\,\frac{\partial\mathfrak{N}_{ph}}{\partial S} - \rho\,\frac{\partial\mathfrak{N}_{ph}}{\partial\rho} \tag{21-19}$$

$$\frac{\partial\mu}{\partial\mu_r} = -\frac{\partial\mathfrak{N}_r}{\partial\rho} \qquad \frac{\partial\mu}{\partial\mu_{ph}} = -\frac{\partial\mathfrak{N}_{ph}}{\partial\rho} \tag{21-20}$$

By using Eqs. 21-22 and 21-23, we can rewrite the expressions for the second viscosity coefficients in helium II in the following form:

$$\zeta_1 = -\frac{1}{\Gamma_{ph}}\,\frac{\partial\mathfrak{N}}{\partial\rho}\left(\mathfrak{N} - \frac{\partial\mathfrak{N}}{\partial S}\,S - \frac{\partial\mathfrak{N}}{\partial\rho}\,\rho\right) - \frac{1}{\Gamma_{r\,ph}}\,\frac{\partial\mathfrak{N}_r}{\partial\rho} \times \left(\mathfrak{N}_r - \frac{\partial\mathfrak{N}_r}{\partial S}\,S - \frac{\partial\mathfrak{N}_r}{\partial\rho}\,\rho\right) \tag{21-21}$$

$$\zeta_2 = \frac{1}{\Gamma_{ph}}\left(\mathfrak{N} - \frac{\partial\mathfrak{N}}{\partial S}\,S - \frac{\partial\mathfrak{N}}{\partial\rho}\,\rho\right)^2 + \frac{1}{\Gamma_{r\,ph}} \times \left(\mathfrak{N}_r - \frac{\partial\mathfrak{N}_r}{\partial S}\,S - \frac{\partial\mathfrak{N}_r}{\partial\rho}\,\rho\right)^2 \tag{21-22}$$

$$\zeta_3 = \frac{1}{\Gamma_{ph}}\left(\frac{\partial\mathfrak{N}}{\partial\rho}\right)^2 + \frac{1}{\Gamma_{r\,ph}}\left(\frac{\partial\mathfrak{N}_r}{\partial\rho}\right)^2 \tag{21-23}$$

$$\zeta_4 = \zeta_1 \tag{21-24}$$

The coefficients ζ_1 and ζ_4 turned out to be equal, as was to be expected from the symmetry principle for kinetic coefficients. The derivatives entering into the expressions for the second viscosity coefficients can be calculated from the known expressions for the number of phonons and rotons and the entropy of helium II.

The absorption of first sound in helium II is determined by the magnitude of the coefficient ζ_2 (cf. Chap. 12). The experimental values of the coefficient of absorption of first sound in helium II in the temperature range from 1.57 to 2°K are sufficiently accurate to permit a determination of the unknown coefficients a and b in the formulas for Γ_{ph} and $\Gamma_{r\,ph}$. In this manner we find

$$\Gamma_{ph} \approx 1.10^{43}\, T^{11} \qquad \Gamma_{r\,ph} \approx 4.10^{50}\, e^{2\Delta/T} \tag{21-25}$$

The value thus obtained for $\Gamma_{r\,ph}$ naturally turns out to be less than the upper bound given in Eq. 7-47.

22

SOUND IN HELIUM II NEAR ZERO TEMPERATURE

Let us investigate the phenomenon of sound propagation in helium II near zero temperature.

Near the absolute zero the mean free path of the excitations increases rapidly and may be larger than the wavelength of sound. In this temperature range only the phonons play a significant role and the rotons may be neglected. Obviously, under these conditions the equations of hydrodynamics for the excitation gas are not applicable any more. The properties of such a gas may be described by the kinetic equation, in which the collision integral is negligibly small. Thus the phonon distribution function satisfies the equation

$$\frac{\partial n}{\partial t} + \frac{\partial n}{\partial \mathbf{r}} \cdot \frac{\partial H}{\partial \mathbf{p}} - \frac{\partial n}{\partial \mathbf{p}} \cdot \frac{\partial H}{\partial \mathbf{r}} = 0 \qquad (22\text{-}1)$$

Here $H = \epsilon(p) + \mathbf{p} \cdot \mathbf{v}_s$, and ϵ is the energy of the phonons which depends on the density; taking the dispersion of the phonons into account we shall write ϵ in the form

$$\epsilon(p) = cp(1 - \gamma p^2) \qquad (22\text{-}2)$$

The quantities ρ and $\mathbf{v}_s$, which play the role of external conditions for the phonons, are determined by the two equations (cf. Eqs. 18-13 and 18-8)

$$\frac{\partial \rho}{\partial t} + \operatorname{div}(\rho \mathbf{v}_s + \int \mathbf{p} n \, d\tau_{\mathbf{p}}) = 0 \qquad (22\text{-}3)$$

$$\frac{\partial \mathbf{v}_s}{\partial t} + \nabla\left(\mu_0 + \frac{\mathbf{v}_s^2}{2} + \int \frac{\partial \epsilon}{\partial \rho}\, n\, d\tau_{\mathbf{p}}\right) = 0 \tag{22-4}$$

where μ_0 is the value of the chemical potential at absolute zero, and $\mathbf{p}$ is the momentum of the excitation in the frame of reference in which $\mathbf{v}_s = 0$. Let us set $n = n_0 + n'$, $\rho = \rho_0 + \rho'$; n_0 and ρ_0 are equilibrium values and n', ρ', and $\mathbf{v}_s$ small quantities, varying as $\exp(-i\omega t + i\mathbf{k}\cdot\mathbf{r})$. Linearizing Eqs. 22-1, 22-3 and 22-4 we obtain the system

$$(\omega - \mathbf{k}\cdot\mathbf{v})n' + \mathbf{v}\cdot\mathbf{k}\frac{\partial n_0}{\partial \epsilon}\left(\frac{\partial \epsilon}{\partial \rho}\rho' + \mathbf{p}\cdot\mathbf{v}_s\right) = 0$$

$$\omega\rho' - \mathbf{k}\rho\cdot\mathbf{v}_s - \mathbf{k}\cdot\int \mathbf{p}n'\, d\tau = 0 \tag{22-5}$$

$$-\omega\mathbf{v}_s + \mathbf{k}\left(\frac{c^2}{\rho} + \int n_0 \frac{\partial^2 \epsilon}{\partial \rho^2}\, d\tau\right)\rho' + \mathbf{k}\int \frac{\partial \epsilon}{\partial \rho}\, n'\, d\tau = 0$$

Here $\mathbf{v} = \partial\epsilon/\partial\mathbf{p}$, and we have used the fact that, in accordance with Eq. 8-26,† $d\mu_0 = dp_0/\rho = c^2(\partial\rho/\rho)$.

Let us denote by θ the angle between the vectors $\mathbf{k}$ and $\mathbf{p}$ and solve the first of Eqs. 22-5 for n'

$$n' = -\frac{\partial n_0}{\partial \epsilon}\, p\, \frac{v\cos\theta}{\omega/k - v\cos\theta}\left(\frac{\partial c}{\partial \rho}\rho' + v_s\cos\theta\right) \tag{22-6}$$

Inserting n' into the other two equations in Eq. 22-5 and integrating over θ we find

$$\left[-\frac{\omega}{k} - \frac{\partial c}{\partial \rho}\int_0^\infty \frac{p^4\, dp}{4\pi^2\hbar^3}\frac{\partial n_0}{\partial \epsilon}\left(\frac{\omega}{kv}\right)^2 \ln\frac{\omega + kv}{\omega - kv}\right]v_s + \left[\frac{c^2}{\rho} - \left(\frac{\partial c}{\partial \rho}\right)\int_0^\infty \frac{p^4\, dp}{4\pi^2\hbar^3}\frac{\partial n_0}{\partial \epsilon}\left(\frac{\omega}{kv}\right)\ln\frac{\omega + kv}{\omega - kv}\right]\rho' = 0$$

$$\left[\rho - \int_0^\infty \frac{p^4\, dp}{4\pi^2\hbar^3}\frac{\partial n_0}{\partial \epsilon}\left(\frac{\omega}{kv}\right)^3 \ln\frac{\omega + kv}{\omega - kv}\right]v_s - \left[\frac{\omega}{k} + \frac{\partial c}{\partial \rho}\int_0^\infty \frac{p^4\, dp}{4\pi^2\hbar^3}\frac{\partial n_0}{\partial \epsilon}\left(\frac{\omega}{kv}\right)^2 \ln\frac{\omega + kv}{\omega - kv}\right]\rho' = 0 \tag{22-7}$$

---

†In general, just as for Fermi liquids, there exists a functional dependence of the excitation energy on the density of excitations of the form $\delta\epsilon = \int f\,\delta n\, d\tau$. However, in the present case this effect is small. Its contribution to our results will be of order $f N_{ph}/kT$. A simple estimate shows that $f N_{ph}/kT \sim (\tau kT/\hbar)^{-1/2} \ll 1$ since $\omega\tau \gg 1$ and $kT/\hbar \gg \omega$.

Here we keep only the large logarithmic terms; the logarithm turns out to be large since $\omega \approx kc$ and $v \approx c$.

The condition of compatibility of the system, Eq. 22-7, yields the dispersion equation

$$\frac{\omega}{k}^{2} - c^{3} = -\int_0^\infty \frac{p^4\, dp}{4\pi^2\, \hbar^3} \frac{\partial n_0}{\partial \epsilon} \ln \frac{\omega + kv}{\omega - kv}\left(\frac{\rho}{c}\frac{\partial c}{\partial \rho} + 1\right)^2 \qquad (22\text{-}8)$$

At low temperatures we perform the simple integration and obtain

$$\delta c = \frac{\omega}{k} - c = \frac{\pi^2}{30\hbar^3 \rho}\left(\frac{kT}{c}\right)^4\left(\frac{\rho}{c}\frac{\partial c}{\partial \rho} + 1\right)^2 \ln\left[\frac{2}{27\gamma}\left(\frac{c}{kT}\right)^2\right] \qquad (22\text{-}9)$$

Thus the velocity of sound at low temperatures is greater than at absolute zero by the amount δc given in Eq. 22-9.

23

HEAT EXCHANGE BETWEEN A SOLID AND HELIUM II

The phenomenon of heat transfer between a solid body and liquid helium has a number of specific features, due to the large thermal resistance of the surface separating these two media. This large thermal resistance leads to the appearance of a temperature jump at such a boundary, whenever there is a flux of heat from one of the media to the other. The magnitude of this temperature jump is proportional to the heat flux and varies with temperature as T^{-3}.

The reason why heat transfer is difficult between a solid body and helium II, may be qualitatively understood by noting the considerable difference in the velocities of sound in helium II and in solids (the velocity of sound in helium II is one order of magnitude less than that in solids). Because of this difference the momenta of phonons in a solid and in liquid helium at the same temperature, will be very different. This means that phonons will not be able to go from the solid to the helium (or conversely), since it is impossible to satisfy simultaneously the laws of conservation of momentum and energy for arbitrary angles of incidence.†

Let us consider a unit area of a boundary surface between a solid and helium, through which there flows a flux Q. In order to find the relation between Q and the magnitude of the temperature jump δT, we shall consider that the helium is in equilibrium at the temperature T and the solid at the temperature $T + \delta T$, and calculate the heat flux from the solid into the helium.

Let us choose a system of coordinates such that the z-axis is

†For the same reasons it will be even more difficult for rotons in the helium to be transformed into phonons in the solid.

normal to the limiting surface, and the region $z > 0$ is filled with liquid helium. We consider the reflection of a plane sound wave coming from the liquid. The velocity field in the liquid may be determined by a scalar potential φ, such that

$$\mathbf{v} = \text{grad}\ \varphi \tag{23-1}$$

In the solid we may represent the displacement vector $\mathbf{u}$ in terms of a scalar Φ and a vector $\mathbf{\Psi}$ by the relation

$$\mathbf{u} = \text{grad}\ \Phi + \text{rot}\ \mathbf{\Psi} \tag{23-2}$$

Let the incident wave vector $\mathbf{k}$ lie in the xz-plane. We choose the potential $\mathbf{\Psi}$ such that $\Psi_x = \Psi_z = 0$ and write $\Psi_y = \psi$. Furthermore let the angle of incidence of the sound wave be θ and its frequency ω. We then have

$$\varphi = A_0 e^{ik(x \sin\theta - z\cos\theta) - i\omega t} + A e^{ik(x\sin\theta + z\cos\theta) - i\omega t}$$

$$\Phi = A_\ell e^{ik_\ell(x\sin\theta_\ell - z\cos\theta_\ell) - i\omega t} \tag{23-3}$$

$$\psi = A_t e^{ik_t(x\sin\theta_t - z\cos\theta_t) - i\omega t}$$

where $k = \omega/c$, $k_\ell = \omega/c_\ell$, $k_t = \omega/c_t$, and because of the homogeneity of the problem in the x-direction we have

$$\frac{\sin\theta}{c} = \frac{\sin\theta_t}{c_t} = \frac{\sin\theta_\ell}{c_\ell}$$

Here c is the velocity of sound in the liquid, c_ℓ, c_t are the longitudinal and transverse sound velocities in the solid. For simplicity we suppose that the solid is isotropic.

Since for $\psi = 0$, we have curl $\mathbf{u} = 0$ and for $\Phi = 0$, div $\mathbf{u} = 0$, we see that A_ℓ and A_t are, respectively, the amplitudes of longitudinal and transverse waves in the solid.

On the limiting surface—that is, for $z = 0$, the normal components of the displacement and stress should be continuous. The first condition may be written in the form

$$-k_\ell \cos\theta_\ell A_\ell + k_t \sin\theta_t A_t = k\cos\theta\,(A - A_0) \tag{23-4}$$

Since in the liquid the stress tensor is equal to $\sigma_{ik} = \ p\ \delta_{ik}$, where the pressure $p = -\rho\ \partial\varphi/\partial t$, and in the solid we have

$$\sigma_{ik} = 2\mathfrak{D}c_t^2 u_{ik} + \mathfrak{D}(c_\ell^2 - 2c_t^2)u_{\ell\ell}\delta_{ik}$$
$$u_{ik} = \frac{1}{2}\left(\frac{\partial u_i}{\partial r_k} + \frac{\partial u_k}{\partial r_i}\right) \tag{23-5}$$

we may write the condition of continuity of the stresses as

$$k_t^2 \cos 2\theta_t \cdot A_t + k_\ell^2 \sin 2\theta_\ell \cdot A_\ell = 0 \tag{23-6}$$

$$\left(1 - 2\frac{c_t^2}{c_\ell^2}\sin^2\theta_\ell\right)A_\ell - \sin 2\theta_t \cdot A_t = \frac{\rho}{\mathfrak{D}}(A + A_0) \tag{23-7}$$

where ρ is the density of liquid and $\mathfrak{D}$ the density of the solid.

For a given value of A_0 the system, Eqs. 23-5 to 23-7, has the following solution:

$$\frac{A}{A_0} = \frac{Z_\ell \cos^2 2\theta_t + Z_t \sin^2 2\theta_t - Z}{Z_\ell \cos^2 2\theta_t + Z_t \sin^2 2\theta_t + Z} \tag{23-8}$$

$$\frac{A_\ell}{A_0} = \frac{\rho}{\mathfrak{D}}\,\frac{2Z_\ell \cos 2\theta_t}{Z_\ell \cos^2 2\theta_t + Z_t \sin^2 2\theta_t + Z} \tag{23-9}$$

$$\frac{A_t}{A_0} = -\frac{\rho}{\mathfrak{D}}\,\frac{2Z_t \sin 2\theta_t}{Z_\ell \cos^2 2\theta_t + Z_t \sin^2 2\theta_t + Z} \tag{23-10}$$

where

$$Z_\ell = \frac{\mathfrak{D}c_\ell}{\cos\theta_\ell} \qquad Z_t = \frac{\mathfrak{D}c_t}{\cos\theta_t} \qquad Z = \frac{\rho c}{\cos\theta}$$

Let us now write down an expression for the transmission coefficient w of a sound wave, defined as the ratio of the energy per unit time going into the solid to the energy incident on the metal per unit time. The numerator is equal to the work of the pressure forces in the liquid—that is,

$$\overline{pv_z} = i\omega\rho\overline{\varphi\frac{\partial\varphi}{\partial z}} \tag{23-11}$$

and the denominator

$$\frac{\rho k^2}{2} c \cos\theta |A_0|^2$$

In Eq. 23-11 the quantity φ is understood to mean the real part of the corresponding complex quantity given in Eq. 23-3, so that Eq. 23-11 may be transformed to

$$\frac{\omega\rho}{4} \overline{(\varphi + \varphi^*)\left(i\frac{\partial\varphi}{\partial z} - i\frac{\partial\varphi}{\partial z}\right)} = \frac{\omega\rho}{2} \operatorname{Re}\left(i\frac{\partial\varphi}{\partial z}\varphi^*\right)$$

If we now insert Eq. 23-3 we find

$$\omega = \operatorname{Re}\left\{\left(1 - \frac{A}{A_0}\right)\left(1 + \frac{A^*}{A_0^*}\right)\right\} \tag{23-12}$$

We may apply Eqs. 23-12 and 23-8 to the calculation of the heat flux W from the liquid to the solid. If we note that when a phonon of energy $\hbar\omega$ is incident on the boundary, an energy $\hbar\omega w$ is transmitted to the solid, we find

$$W = \int n\left(\frac{\hbar\omega}{kT}\right) c\cdot\cos\theta\ \hbar\omega\cdot w\ \frac{dk}{(2\pi)^3} \tag{23-13}$$

where the angular integration is to be carried out over the interval $0 \leqslant \cos\theta \leqslant 1$ and n is the Planck function for the phonons. By using Eq. 23-12 we obtain

$$W = \operatorname{Re} \int_0^\infty n\left(\frac{\hbar\omega}{kT}\right) \hbar\omega\cdot c\cdot\frac{2\pi\omega^2\, d\omega}{(2\pi c)^3}$$

$$\times \int_0^1 \cos\theta \left(1 - \frac{A}{A_0}\right)\left(1 + \frac{A^*}{A_0^*}\right) d(\cos\theta) \tag{23-14}$$

The frequency integral may be performed quite simply, and as a result we get

$$W = \frac{\rho}{\mathfrak{D}} c \frac{4\pi^5}{15} \frac{T^4}{(2\pi\hbar c_t)^3} F \tag{23-15}$$

where

$$F = \frac{1}{2}\frac{\mathfrak{D}}{\rho}\left(\frac{c_t}{c}\right)^3 \operatorname{Re} \int_0^1 \cos\theta \left(1 - \frac{A}{A_0}\right)\left(1 + \frac{A^*}{A_0^*}\right) d(\cos\theta) \tag{23-16}$$

The integral occurring in Eq. 23-16 may be divided into three parts

$$\mathcal{g}_1 = \int_0^{\sqrt{1-c^2/c_t^2}} \cos\theta \left(1 - \frac{A}{A_0}\right)\left(1 + \frac{A^*}{A_0^*}\right) d(\cos\theta) \qquad (23\text{-}17)$$

$$\mathcal{g}_2 = \int_{\sqrt{1-c^2/c_t^2}}^{\sqrt{1-c^2/c_\ell^2}} \cos\theta \left(1 - \frac{A}{A_0}\right)\left(1 + \frac{A^*}{A_0^*}\right) d(\cos\theta) \qquad (23\text{-}18)$$

$$\mathcal{g}_3 = \int_{\sqrt{1-c^2/c_\ell^2}}^{1} \cos\theta \left(1 - \frac{A}{A_0}\right)\left(1 - \frac{A^*}{A_0^*}\right) (d\cos)\theta \qquad (23\text{-}19)$$

We insert the value of A/A_0 from Eq. 23-8 and note that in the interval $0 \leqslant \cos\theta \leqslant \sqrt{1 - c^2/c_t^2}$ the quantities $\cos\theta_\ell$ and $\cos\theta_t$ are purely imaginary (we have $\cos\theta_{\ell,t} = i|\cos\theta_{\ell,t}|$), and in the interval $\sqrt{1 - c^2/c_t^2} \leqslant \cos\theta \leqslant \sqrt{1 - c^2/c_\ell^2}$, $\cos\theta_t$ is real and $\cos\theta_\ell$ is pure imaginary; we thus have

$$\mathcal{g}_1 = 4\rho c \int_0^{\sqrt{1-c^2/c_t^2}} \frac{Z_\ell \cos^2 2\theta_t + Z_t \sin^2 2\theta_t}{(Z_\ell \cos^2 2\theta_t + Z_t \sin^2 2\theta_t)^2 - Z^2}\, d(\cos\theta) \qquad (23\text{-}20)$$

$$\mathcal{g}_2 = 4\rho c \int_{\sqrt{1-c^2/c_\ell^2}}^{\sqrt{1-c^2/c_\ell^2}} \frac{Z_t \sin^2 2\theta_t}{|Z_\ell|^2 \cos^4 2\theta_t + Z_t^2 \sin^4 2\theta_t}\, d(\cos\theta) \qquad (23\text{-}21)$$

$$\mathcal{g}_3 = 4\rho c \int_{\sqrt{1-c^2/c_\ell^2}}^{1} \frac{Z_\ell \cos^2 2\theta_t + Z_t \sin^2 2\theta_t}{(Z_\ell \cos^2 2\theta_t + Z_t \sin^2 2\theta_t)^2}\, d(\cos\theta) \qquad (23\text{-}22)$$

In the denominators of the integrands of Eqs. 23-21 and 23-22 we have neglected Z in comparison with $Z_{\ell,t}$, since the ratio $\rho c/\mathfrak{D}c_t$ is exceedingly small.

Let us consider the first integral, $\mathcal{g}_1$, and integrate formally along the real axis of the complex $\cos\theta$ plane. Since the functions Z_ℓ and Z_t are pure imaginary, we see that the whole integral is imaginary and does not contribute to Eq. 23-16. This argument, however, omits one important point, namely that the integrand of $\mathcal{J}_1$ has a pole in the upper half-plane, which is extremely close to the real value $\cos\theta = \cos\theta_1$, where $\cos\theta_1$ is determined by the condition

$$Z_\ell \cos^2 2\theta_t + Z_t \sin^2 2\theta_t - Z = 0 \tag{23-23}$$

This is the condition for surface waves on the free surface of the solid (Raleigh waves). From Eq. 23-8 we see that it is also the condition that the amplitude of the reflected wave should vanish. Thus the pole corresponds to a solution representing a nearly plane wave in the liquid incident on the surface, and a wave in the solid which is nearly a Raleigh wave. The distance from the pole to the real axis is extremely small (in order of magnitude it is equal to the ratio of the acoustical resistances in the liquid and the solid—that is, $\rho c/\mathfrak{D}c_t \sim 10^{-2}$); therefore a relatively small displacement of the contour of integration into the upper half-plane (such a displacement can in fact always be brought about by the damping which is present) will bring the pole into a position between the contour and the real axis. Thus the integration in Eq. 23-20 must be carried out along the contour c_0 represented in Fig. 6. It is convenient to deform this contour into the contour C. The part of C which is along the real axis does not give any contribution to Eq. 23-16, for the reason mentioned previously. The remaining part may be expressed in the usual way in terms of the residue of the integrand at the pole. This yields

$$\mathcal{J}_1 = -4\pi i\rho c \left\{\frac{d}{d\cos\theta}\left[Z_\ell \cos^2 2\theta_t + Z_t \sin^2 2\theta_t\right]_{\cos\theta = \cos\theta_1}\right\}^{-1} \tag{23-24}$$

If we carry out the differentiation in Eq. 23-24 and set $\theta = \theta_1$, we obtain

$$\mathcal{J}_1 = \frac{\pi\rho c^3}{2Dc_t^3}\left\{(2\sin^2\theta_t - 1)\right.$$
$$\times\left(\frac{1}{\sqrt{\sin^2\theta_t - 1}} - \frac{1}{\sqrt{\sin^2\theta_t - c_t^2/c_\ell^2}}\right) + \sqrt{\sin^2\theta_t - 1}$$
$$\left.\times\frac{2\sin^2\theta_t - c_t^2/c_\ell^2}{2(\sin^2\theta_t - c_t^2/c_\ell^2)}\right\}^{-1}_{\theta=\theta_1} \tag{23-25}$$

where we took into account the smallness of c^2/c_t^2.

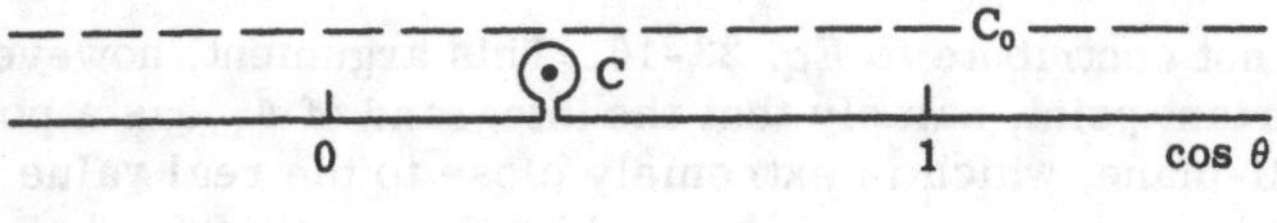

Figure 6.

Let us introduce the notation

$$\xi\left(\frac{c_t}{c_\ell}\right) = \frac{1}{\sin\theta_t}\bigg|_{\theta=\theta_1} \qquad \eta = \frac{c_\ell}{c_t}$$

and rewrite Eq. 23-25 in the form

$$\mathcal{J}_1 = \frac{\pi\rho c^3}{2\mathfrak{D}c_t^3}\left\{\left(\frac{2}{\xi}-\xi\right)\left(\frac{1}{2\sqrt{1-\xi^2}}-\frac{1}{\sqrt{1-\xi^2/\eta^2}}\right) + \frac{\sqrt{1-\xi^2}}{\xi}\,\frac{2-\xi^2/\eta^2}{2(1-\xi^2/\eta^2)}\right\}^{-1} \tag{23-26}$$

We now consider the integral $\mathcal{J}_2$ defined by Eq. 23-21. If we make a change of variable of integration from $\cos\theta$ to $\cos\theta_t$ we obtain

$$\mathcal{J}_2 = \frac{2\rho c^3}{\mathfrak{D}}\cdot\frac{1}{2c_t^3}\int_0^{\sqrt{1-c_t^2/c_\ell^2}} d(\cos\theta_t) \times \frac{4c_t^4\cos^2\theta_t\,|\sin^2 2\theta_\ell|}{c_\ell^4\cos^4 2\theta_t + c_t^4\sin^2 2\theta_t\,|\sin^2 2\theta_\ell|} \tag{23-27}$$

In order to discuss the integral $\mathcal{J}_3$ given in Eq. 23-22 we divide it into two parts

$$\mathcal{J}_3 = 4\rho c\int_{\sqrt{1-c^2/c_\ell^2}}^{1} d(\cos\theta)\,\frac{Z_\ell\cos^2 2\theta_t}{(Z_\ell\cos^2 2\theta_t + Z_t\sin^2 2\theta_t)^2}$$

$$+\; 4\rho c\int_{\sqrt{1-c^2/c_\ell^2}}^{1} d(\cos\theta)\,\frac{Z_t\sin^2 2\theta_t}{(Z_\ell\cos^2 2\theta_t + Z_t\sin^2 2\theta_t)^2} \tag{23-28}$$

In the first part we make a change of variable to $\cos\theta_\ell$ and in the second part to $\cos\theta_t$. Then $\mathcal{I}_3$ may be written as

$$\mathcal{I}_3 = \frac{2\rho c^3}{\mathcal{D}}\left\{\frac{1}{2c_t^3}\int_{\sqrt{1-c_t^2/c_\ell^2}}^{1}\frac{4c_t^4\cos^2\theta_t\sin^2\theta_\ell}{(c_\ell^2\cos^2 2\theta_t + c_t^2\sin 2\theta_t\sin 2\theta_\ell)^2}\,d(\cos\theta_t)\right.$$

$$\left. + \frac{1}{2c_\ell^3}\int_0^1 d(\cos\theta_\ell)\frac{4c_\ell^4\cos^2\theta_\ell\cos^2 2\theta_t}{(c_t^2\sin 2\theta_\ell\sin 2\theta_t + c_\ell^2\cos^2 2\theta_t)^2}\right\}$$

(23-29)

The sum $\mathcal{I}_2 + \mathcal{I}_3$ may be somewhat simplified by introducing the variables of integration $y = \sin^2\theta_t$ in Eq. 23-27 and in the first integral in Eq. 23-29, and $y = \sin^2 2\theta_t$ in the second integral in Eq. 23-29. This yields

$$\mathcal{I}_2 + \mathcal{I}_3 = \frac{4\rho c^3}{\mathcal{D}c_t^3}\left\{\int_0^{1/\eta^2}\frac{\sqrt{1-y\eta^2}\,dy}{\eta(1-2y^2) + 4y\sqrt{(1-y)(1-y\eta^2)}}\right.$$

$$\left. + \int_{1/\eta^2}^{1}\frac{4y(y\eta^2-1)\,dy}{(1-2y^2)^4\eta^2 + 16y^2(1-y)(y\eta^2-1)}\right\}$$

(23-30)

By some simple manipulations this may be transformed to

$$\mathcal{I}_2 + \mathcal{I}_3 = \frac{4\rho c^3}{\mathcal{D}c_t^3}\left\{\int_0^1\frac{4y\sqrt{1-y}\,(y\eta^2-1)\,dy}{\eta^2(1-2y^2)^2 + 16y^2(1-y)(y\eta^2-1)}\right.$$

$$\left. + \int_0^{1/\eta^2}\frac{\eta\sqrt{1-y\eta^2}\,(1-2y)^2\,dy}{\eta^2(1-2y)^2 + 16y^2(1-y)(y\eta^2-1)}\right\}$$

(23-31)

Finally, if we insert Eqs. 23-31 and 23-26 into Eq. 23-16 we recover formula 23-15 for the heat flux from the liquid to the solid W, with the explicit formula

$$F = 2\int_0^1 \frac{4y\sqrt{1-y}\,(y\eta^2-1)\,dy}{\eta^2(1-2y)^4 + 16y^2(1-y)(y\eta^2-1)}$$

$$+ 2\int_0^{1/\eta^2} \frac{\eta\sqrt{1-y\eta^2}\,(1-2y)\,dy}{\eta^2(1-2y)^4 + 16y^2(1-y)(y\eta^2-1)} + \frac{\pi}{4}\left\{\left(\frac{2}{\xi} - \xi\right)\right.$$

$$\left.\times\left(\frac{1}{2\sqrt{1-\xi^2}} - \frac{\eta}{\sqrt{1-\xi^2/\eta^2}}\right) + \frac{1}{\xi}\sqrt{1-\xi^2}\,\frac{2-\xi^2/\eta^2}{2(1-\xi^2/\eta^2)}\right\}^{-1} \tag{23-32}$$

The numerical values of F depend on the elastic constants of the solid; nevertheless F varies within rather narrow limits, always remaining of order unity.

If the temperature T of the liquid is equal to the temperature of the solid, the heat flux W is compensated by another current, which is equal to W in magnitude, but directed from the solid to the liquid. If the temperature difference δT is small then the resulting heat flux Q is equal to

$$Q = \frac{\partial W}{\partial T}\,\delta T$$

Let us carry out the differentiation in Eq. 23-15, thus obtaining the final relation between the temperature jump δT and the heat flux through the boundary

$$Q = \frac{\rho c}{\mathfrak{D}}\,\frac{16\pi^5}{15}\,\frac{T^3\,\delta T}{(2\pi\hbar c_t)^3}\,F \tag{23-33}$$

Thus the thermal resistivity of the boundary is proportional to T^{-3}—that is, it grows rather rapidly with decreasing temperature.

PART IV

IMPURITIES IN HELIUM II

24

SOLUTIONS OF IMPURITIES IN HELIUM II

The properties of weak solutions of impurities in superfluid helium may be discussed on the basis of extremely general considerations. By impurities we will mean atoms of the isotope He^3, electrons, ions of He^4, etc. In weak solutions the solute particles will practically not interact with one another. Since they have a large wavelength they will not be localized in one spot in the liquid. The interaction of the impurities with the atoms of He^4 leads to the appearance of additional energy levels. The states of the impurities may be classified by the values of the momentum, which is a continuous variable. Thus to each impurity there corresponds an elementary excitation characterized by an energy which is a function of the momentum. In principle various types of spectra are possible. However, in practice, we shall always deal with the simplest case in which the energy ϵ is just a quadratic function of the momentum $\mathbf{p}$ with an effective mass m^*

$$\epsilon = \epsilon_0 + \frac{p^2}{2m^*} \qquad (24\text{-}1)$$

(ϵ_0 is some constant "zero-point energy").

Thus when the impurities are atoms of He^3, these have a spectrum of type Eq. 24-1 with an effective mass equal to $m^* = 2.8 m_{He^3}$. It appears that electrons and ions also have a spectrum of the form Eq. 24-1. As long as the velocity of the impurity is less than the velocity of sound in superfluid helium, the impurity is unable to emit phonons. The emission of rotons is also impossible as long as the energy $p^2/2m^*$ of the impurity is less than the roton energy Δ. Thus impurities moving through the liquid with "subsonic" velocities, will not interact with the superfluid part. However they will collide and interact with the phonons and rotons and will, of course, be carried along by their motion. Thus for weak solutions the impurities will participate in the normal motion of the liquid. It is important to emphasize that this conclusion that the impurities take part in the normal motion, is independent of whether or not the impurities themselves can be superfluid. For example, atoms of the short-lived isotope He^6 are capable of forming a superfluid; however, in weak solution in He^4 they will, nevertheless, participate only in the normal motion. For temperatures which are not too low, the distribution of impurity atoms in energy will be determined by Boltzmann statistics. For very low temperatures it is necessary to take into account the degeneracy of the impurity gas and the interaction of the impurity atoms with one another. If the spin s of the particles is a half-integer, the degeneracy temperature T_0 is equal to

$$T_0 = \frac{\hbar^2 N_0^{2/3} c^{2/3}}{km^*}\left(\frac{3\pi^2}{\sqrt{2}(2s+1)}\right)^{2/3}$$

For impurities having integer spin we have

$$T_0 = \frac{\hbar^2 N_0^{2/3} c^{2/3}}{km^*}\,\frac{\pi 3^{2/3}}{2(2s+1)^{2/3}} \tag{24-3}$$

Here $N_0 = \rho/m$ is the number of atoms of He^4 per cm^3, m is the mass of a He^4 atom, and c the concentration of impurities.

When the impurity is an atom of He^3, for $c < 10^{-2}$ we have $T_0 <$ 0.1°K. For low temperatures the interaction of impurity atoms with one another also becomes important. The energy of such an interaction is equal Vc, where the characteristic energy V is equal to a few degrees in the case of He^3.

In the classical region (for $T > T_0$) it is easy to obtain the thermodynamic functions for weak solutions. Since the impurities take part in the normal motion they give a contribution ρ_{ni} to the normal density; this may be calculated by the general formula, Eq. 2-22†

$$\rho_{ni} = \tfrac{1}{3}\int p^2 \frac{\partial n}{\partial \epsilon}\, d\tau$$

Inserting the Boltzmann distribution function

†The calculations that follow will be carried out for the physically interesting case in which the spectrum has the form Eq. 24-1.

$$n = N_0 c\,(2\pi m^* kT)^{-2/3}\, e^{-p^2/2m^* kT} \tag{24-4}$$

we find the obvious result

$$\rho_{ni} = \rho c m^*/m \tag{24-5}$$

For sufficiently low temperatures the contribution of the phonons and rotons to the normal density, ρ_{no}, decreases rapidly, and the constant impurity contribution ρ_{ni} may become larger than ρ_{no}. For concentrations of the order of 10^{-10} we have $\rho_{no} \approx \rho_{ni}$ for temperatures of about 1°K.

THE ENTROPY

If we apply the general formulas of the thermodynamics of weak solutions we find the entropy of 1 cm³ of solution the following expression:

$$S = S_0 + kN_0 c \ln\left[\frac{(2s+1)}{N_0 c}\left(\frac{m^* kT}{2\pi\hbar^2}\right)^{3/2} + \frac{5}{2}\right] \tag{24-6}$$

where S_0 is the entropy of pure helium.

THE SPECIFIC HEAT

The specific heat of 1 cm³ of helium containing impurities is equal to

$$C = C_0 + \tfrac{3}{2} N_0 ck \tag{24-7}$$

where C_0 is the specific heat of pure helium.

The contribution of impurities is less noticeable in the case of the specific heat than for the normal density or the entropy.

THE EQUATIONS OF HYDRODYNAMICS FOR SOLUTIONS

The basic property of solutions which will be used to derive the equations of hydrodynamics is the fact that the impurities are carried along completely with the normal motion. For weak solutions this assertion has been proven rigorously. A number of arguments may be given to show that even in concentrated solutions this is the case, namely, that at all concentrations the impurities only participate in the normal motion. There does not exist any rigorous theoretical demonstration of this assertion. However in the most interesting case, when the impurities are atoms of He^3, it may be verified experimentally, by measuring the normal density of the solution. At this point it is necessary to state a number of reservations. The above statements hold only if the impurities themselves cannot form a superfluid. In He^3, at sufficiently low temperatures (below 0.01°K) it appears that a transition is possible to a paired state with nonzero angular momentum, which would have superfluid properties. A similar pairing phenomenon is possible also in solutions, at an even lower temperature. Therefore our statement that the impurities participate only in the normal motion cannot be extended down to regions of temperatures at which the impurities may become superfluid.†

Moreover, for solutions of He^3 in He^4 there is a region in the phase diagram below 0.8°K, in which the system breaks up into two phases. We shall be concerned with solutions outside this region.

Thus, assuming that the impurities participate only in the normal motion, we have, along with the equation of continuity,

$$\frac{\partial}{\partial t}\rho + \operatorname{div} \mathbf{j} = 0 \tag{24-8}$$

which expresses the conservation of matter for the whole liquid, the equation of continuity for the impurities

$$\frac{\partial}{\partial t}\rho c + \operatorname{div} \rho c \mathbf{v}_n = 0 \tag{24-9}$$

Equation 24-9 is the mass conservation law for the impurities, which are transported with the velocity $\mathbf{v}_n$.

Let us further write down the continuity equation for entropy, taking into account the fact that the entropy is also transported with the normal motion

$$\dot{S} + \operatorname{div} S\mathbf{v}_n = 0 \tag{24-10}$$

The equation of motion

$$\frac{\partial j_i}{\partial t} + \frac{\partial \Pi_{ik}}{\partial r_k} = 0 \qquad \Pi_{ik} = \rho_s v_{si} v_{sk} + \rho_n v_{ni} v_{nk} + p\,\delta_{ik} \tag{24-11}$$

---

†For mixtures of two superfluids it is in principle possible to realize situations in which three independent motions occur simultaneously: a normal motion with velocity $\mathbf{v}_n$ and two superfluid motions with velocities $\mathbf{v}_s'$ and $\mathbf{v}_s''$. The equations appropriate for such a three-fluid hydrodynamics were written down in Ref. 30.

which expresses the law of conservation of the momentum **j** of the solution, has the same form as for pure helium since the pressure, and consequently also the tensor Π_{ik}, are defined in the same way in the solution as in pure helium.

Finally, remembering that curl $\mathbf{v}_s = 0$, we may write the equation of superfluid motion in the form

$$\dot{\mathbf{v}}_s + \nabla\left(\frac{v_s^2}{2} + \varphi\right) = 0 \qquad (24\text{-}12)$$

where φ is some unknown function, to be determined.

In order to proceed further we use the law of conservation of energy, just as for pure helium. We calculate the time derivative of the total energy and use Eqs. 24-8 to 24-12 to rewrite all time derivatives of thermodynamic variables and velocities, in terms of their spatial derivatives. It must be noted that the thermodynamic identity which determines the internal energy E_0 contains an additional variable, the concentration c

$$dE_0 = T\,dS + \mu\,d\rho + Z\,dc + ((\mathbf{v}_n - \mathbf{v}_s)\cdot d\mathbf{j}_0) \qquad (24\text{-}13)\dagger$$

By grouping the terms thus obtained inside a divergence, we find an expression for the energy flux vector

$$\mathbf{Q} = \mathbf{j}\left(\mu - \frac{Z}{\rho}c + \frac{v_s^2}{2}\right) + Zc\mathbf{v}_n + \mathbf{v}_n(\mathbf{v}_n\cdot\mathbf{j}_0) \qquad (24\text{-}14)$$

From the requirement that the terms that are not expressible as a divergence should vanish, we find the function φ of Eq. 24-12

$$\varphi = \mu - \frac{Z}{\rho}c \qquad (24\text{-}15)$$

The final form of the equation of superfluid motion is therefore

$$\dot{\mathbf{v}}_s + \nabla\left(\frac{v_s^2}{2} + \mu - \frac{Z}{\rho}c\right) = 0 \qquad (24\text{-}16)$$

Let us express the potentials μ and Z which occur in this paragraph in terms of the chemical potential μ_4 of the helium II and the chemical potential μ_3 of the solute particles. The free energy of the solution may be represented in the form

†An expression for the potential Z in terms of the chemical potentials of the solute particles and the helium, will be given below.

$$F = (N_3 m_3 + N_4 m_4) f\left(T, \frac{V}{N_3 m_3 + N_4 m_4}, \frac{N_3 m_3}{N_3 m_3 + N_4 m_4}\right) \tag{24-17}$$

Here N_4 and N_3 are the numbers of helium II atoms and of solute particles respectively, and m_4 and m_3 the corresponding masses. The free energy per unit volume is obviously equal to

$$F = \rho f \tag{24-18}$$

The chemical potential μ_4 of the helium II in the solution is equal to

$$\mu_4 = \frac{\partial F}{m_4 \partial N_4} = f - \frac{\partial f}{\partial v} v - \frac{\partial f}{\partial c} c \tag{24-19}$$

In an analogous manner we find for the solute particles

$$\mu_3 = \frac{\partial F}{m_3 \partial N_3} = f - \frac{\partial f}{\partial v} v + (1 - c)\frac{\partial f}{\partial c} \tag{24-20}$$

for the concentration

$$C = \frac{N_3 m_3}{N_3 m_3 + N_4 m_4}$$

and for the specific volume

$$v = \frac{V}{N_3 m_3 + N_4 m_4} = \frac{1}{\rho}$$

Differentiating Eq. 24-18 and taking into account Eqs. 24-19 and 24-20 we find the desired relation

$$Z = \left(\frac{\partial F}{\partial c}\right)_{\partial, T} = \rho(\mu_3 - \mu_4) \tag{24-21}$$

$$\mu = \left(\frac{\partial F}{\partial \rho}\right)_{T,c} = c\mu_3 + (1 - c)\mu_4 \tag{24-22}$$

We may use Eq. 24-22 to find the entropy per unit mass of solution

$$\sigma = -\left(\frac{\partial \mu}{\partial T}\right)_p = -c\frac{\partial \mu_3}{\partial T} - (1 - c)\frac{\partial \mu_4}{\partial T} \tag{24-23}$$

In the general case, owing to interactions between the solute particles and the helium II, we cannot write down expressions for the chemical potentials μ_3 and μ_4. The only case in which this can be done is the case of ideal solutions, for which we have

$$\mu_3 = \mu_{30} + \frac{kT}{m_3} \ln c \tag{24-24}$$

$$\mu_4 = \mu_{40} + \frac{kT}{m_4} \ln (1 - c) \tag{24-25}$$

(μ_{30} and μ_{40} are the chemical potentials of the pure solute and of pure helium II, respectively). According to Eq. 24-23 the entropy σ may be written

$$\sigma = (1 - c)\sigma_{40} + c\sigma_{30} - \frac{k}{m_4}(1 - c) \ln (1 - c) - \frac{k}{m_3} c \ln c \tag{24-26}$$

By Eqs. 8-21 and 24-13 the thermodynamic potential of the solution satisfies the identity

$$\rho\, d\mu = -S\, dT + dp + Z\, dc - \mathbf{j}_0 \cdot d(\mathbf{v}_n - \mathbf{v}_s) \tag{24-27}$$

For small values of the velocity difference $\mathbf{v}_n - \mathbf{v}_s$ (cf. Chap. 8) this may be integrated to yield

$$\mu = \mu_c(p,T,c) - \frac{\rho_n}{2\rho}(\mathbf{v}_n - \mathbf{v}_s)^2 \tag{24-28}$$

where the potential μ_c satisfies the identity

$$\rho\, d\mu_c = -S\, dT + dp + Z\, dc \tag{24-29}$$

The combination of the potentials μ and Z which occurs in the equation of motion, Eq. 24-26, may according to Eqs. 24-21 and 24-22 be expressed in terms of the chemical potentials by the relation

$$\mu - \frac{Z}{\rho}c = c\mu_3 + (1 - c)\mu_4 - c(\mu_3 - \mu_4) = \mu_4 \tag{24-30}$$

For weak solutions and small values of the difference $\mathbf{v}_n - \mathbf{v}_s$ this combination is by Eq. 24-25 equal to

$$\mu - \frac{Z}{\rho}c = \mu_{40} - \frac{kTc}{m_4} = \mu_0(p,T) - \frac{kTc}{m_4} \tag{24-31}$$

In this case Eq. 24-16 takes the form

$$\dot{\mathbf{v}}_s + \nabla \left\{ \mu_0(p,T) + \frac{v_s^2}{2} - \frac{\rho_n}{2\rho}(\mathbf{v}_n - \mathbf{v}_s)^2 - \frac{kTc}{m_4} \right\} = 0 \qquad (24\text{-}32)$$

From Eqs. 24-11 and 24-16 we find the equilibrium conditions for solutions. In equilibrium, when $\mathbf{v}_n = \mathbf{v}_s = 0$ we have

$$p = \text{const} \qquad (24\text{-}33)$$

$$\mu - \frac{Z}{\rho} c = \text{const} \qquad (24\text{-}34)$$

For weak solutions, Eq. 24-34 may be rewritten in the form

$$\mu_0(p,T) - \frac{kTc}{m_4} = \text{const}$$

which in differential form is $(dp = 0)$

$$-\sigma_0 \, dT - \frac{k}{m_4} d(cT) = 0 \qquad (24\text{-}35)$$

Thus in a solution, in contrast to pure helium II, one may maintain a temperature gradient under equilibrium conditions, and this is automatically accompanied by a concentration gradient. In weak solutions, already for small temperature gradients, almost all the impurities are collected at the cold end.

DISSIPATIVE PROCESSES IN SOLUTIONS

In order to study dissipative processes in solutions we shall proceed in the same way as for pure helium II. In the present case there are additional terms in the equations of motion and in the continuity equation for the solute particles

$$\frac{\partial j_i}{\partial t} + \frac{\partial}{\partial r_k}(\Pi_{ik} + \tau_{ik}) = 0 \qquad (\tau_{ik} = \tau\, \delta_{ik} + \mu_{ik}) \qquad (24\text{-}36)$$

$$\dot{\mathbf{v}}_s + \nabla\left(\mu - \frac{Z}{\rho}c + \frac{v_s^2}{2} + h\right) = 0 \qquad (24\text{-}37)$$

$$\frac{\partial}{\partial t}(\rho c) + \operatorname{div}(\rho c \mathbf{v}_n + \mathbf{g}) = 0 \qquad (24\text{-}38)$$

These additional terms take into account the possible dissipative processes. From the energy conservation law we obtain an expression for the energy flux vector

$$\mathbf{Q} = \mathbf{j}\left(\mu - \frac{Z}{\rho}c + \frac{v_s^2}{2}\right) + Zc\mathbf{v}_n + ST\mathbf{v}_n + \mathbf{v}_n(\mathbf{v}_n \cdot \mathbf{j}_0)$$

$$+ h(\mathbf{j} - \rho\mathbf{v}_n) + \tau\mathbf{v}_n + \mathbf{g} + \nu \qquad (\nu_k = v_{ni}\mu_{ik}) \tag{24-39}$$

and an equation determining the rate of change of the entropy

$$T\left\{\dot{S} + \operatorname{div}\left(S\mathbf{v}_n + \frac{\mathbf{q}}{T} - \frac{\mathbf{g}Z}{\rho T}\right)\right\} = -\left\{h \operatorname{div}(\mathbf{j} - \rho\mathbf{v}_n)\right.$$

$$+ \tau \operatorname{div} \mathbf{v}_n + \tfrac{1}{2}\mu_{ik}\left(\frac{\partial v_{nk}}{\partial n_k} + \frac{\partial v_{nk}}{\partial r_i} - \tfrac{2}{3}\delta_{ik}\frac{\partial v_{n\ell}}{\partial r_\ell}\right)$$

$$\left.\times \frac{\mathbf{q}\cdot\Delta T}{T} + \mathbf{g}T\cdot\nabla\frac{Z}{\rho T}\right\} \tag{24-40}$$

Futhermore, from the requirement that the dissipative function should be positive we conclude that the unknown coefficients have the form

$$\tau = -\zeta_1 \operatorname{div}(\mathbf{j} - \rho\mathbf{v}_n) - \zeta_2 \operatorname{div} \mathbf{v}_n \tag{24-41}$$

$$h = -\zeta_3 \operatorname{div}(\mathbf{j} - \rho\mathbf{v}_n) - \zeta_4 \operatorname{div} \mathbf{v}_n \tag{24-42}$$

$$\mu_{ik} = -\eta \quad \frac{\partial v_{ni}}{\partial r_k} + \frac{\partial v_{nk}}{\partial r_i} - \tfrac{2}{3}\delta_{ik}\frac{\partial v_{n\ell}}{\partial r_\ell} \tag{24-43}$$

$$\mathbf{g} = -\alpha\nabla\frac{Z}{\rho T} - \beta\frac{1}{T^2}\nabla T \tag{24-44}$$

$$\mathbf{q} = -\gamma\nabla\frac{Z}{\rho T} - \delta\frac{1}{T^2}\nabla T \tag{24-45}$$

The symmetry principle for the kinetic coefficients implies the two relations

$$\zeta_1 = \zeta_4 \qquad \beta = \gamma \tag{24-46}$$

The coefficients $\zeta_1, \zeta_2, \zeta_3, \zeta_4$ are coefficients of second viscosity in the solution; η is the coefficient of first viscosity. It is convenient to eliminate the quantity $\nabla(Z/\rho T)$ from the expression for the heat flux $\mathbf{q}$, by expressing it in terms of ∇T and the impurity current $\mathbf{g}$

$$-\mathbf{q} = \frac{\gamma}{\alpha}\mathbf{g} + \left(\delta - \frac{\beta\gamma}{\alpha}\right)\frac{\nabla T}{T^2} \tag{24-47}$$

We define the thermal conductivity such that for zero impurity current $\mathbf{g}$ the heat flux should be equal to $-\kappa\nabla T$; this yields

$$\kappa = \left(\delta - \frac{\beta\gamma}{\alpha}\right)\frac{1}{T^2} \tag{24-48}$$

Furthermore, as is usually done for solutions, we express the impurity current $\mathbf{g}$ in terms of the usual variables p, T, and c and introduce the notation

$$\mathfrak{D} = \frac{\alpha}{\rho}\frac{\partial}{\partial c}\left(\frac{Z}{\rho T}\right) \tag{24-49}$$

$$\rho\mathfrak{D}k_T = \alpha T\frac{\partial}{\partial T}\left(\frac{Z}{\rho T}\right) + \frac{\beta}{T} \tag{24-50}$$

$$k_p = \frac{p\dfrac{\partial}{\partial p}\left(\dfrac{Z}{\rho T}\right)}{\dfrac{\partial}{\partial c}\left(\dfrac{Z}{\rho T}\right)} \tag{24-51}$$

Then the fluxes $\mathbf{g}$ and $\mathbf{q}$ take the form

$$-\mathbf{g} = \rho\mathfrak{D}\left(\nabla c + \frac{k_T}{T}\nabla T + \frac{k_p}{p}\nabla p\right)$$

$$-\mathbf{q} = T^2\left(\frac{\partial}{\partial T}\frac{Z}{\rho T} - \frac{k_T}{T}\frac{\partial}{\partial c}\frac{Z}{\partial T}\right)\mathbf{g} + \kappa\nabla T \tag{24-53}$$

The quantity $\mathfrak{D}$ is the diffusion coefficient, and $k_T\mathfrak{D}$ the coefficient of thermal diffusion. The quantity $k_p\mathfrak{D}$ is called the coefficient of "barodiffusion." It occurs only in the presence of pressure gradients, and does not depend on kinetic properties, but is entirely determined by the thermodynamics of the solution.

Let us investigate the form of the fluxes $\mathbf{g}$ and $\mathbf{q}$ under stationary conditions. According to Eqs. 24-33 and 24-34 we have, under stationary conditions (for small gradients),

$$p = \text{const}$$

$$\mu - \frac{Z}{\rho} c = \text{const}$$

Using the thermodynamic identity given in Eq. 24-27 let us find the relation between the gradients of temperature and concentration under stationary conditions

$$\nabla c \frac{\partial}{\partial c} \frac{Z}{\rho} = -\left[\frac{S}{\rho c} + \frac{\partial}{\partial T}\left(\frac{Z}{\rho}\right)\right] \nabla T = \nabla T \frac{\partial}{\partial c}\left(\frac{S}{\rho c}\right) c \tag{24-54}$$

Under these conditions the total current of solute particles is zero

$$\mathbf{g} + \rho c \mathbf{v}_n = 0 \tag{24-55}$$

and the total heat flux is equal to

$$\mathbf{Q} = (Zc + ST)\mathbf{v}_n + \mathbf{q} \tag{24-56}$$

The velocity $\mathbf{v}_n$ is not equal to zero; from Eq. 24-11 we see only that the total momentum vanishes. Accoring to Eqs. 24-54 and 24-55 both fluxes $\mathbf{g}$ and $\mathbf{q}$ and the velocity $\mathbf{v}_n$ are all proportional to the temperature gradient under stationary conditions. We express these in terms of ∇T and insert them into Eq. 24-56; this yields

$$\mathbf{Q} = -\kappa \nabla T - \rho \mathfrak{D} \frac{T}{\partial/\partial c(Z/\rho)} \left\{c \frac{\partial}{\partial c}\left(\frac{S}{\rho c}\right) + \frac{k_T}{T} \frac{\partial}{\partial c}\left(\frac{Z}{\rho}\right)\right\}^2 \nabla T \tag{24-57}$$

This last equation suggests that we characterize solutions by some effective thermal conductivity κ_{eff} which turns out to be some combination of the diffusion coefficient, the thermal diffusion, and the thermal conductivity; it is equal to

$$\kappa_{eff} = \kappa + \rho \mathfrak{D} \frac{T}{\partial/\partial c(Z/\partial)} \left\{c \frac{\partial}{\partial c}\left(\frac{S}{\rho c}\right) + \frac{k_T}{T} \frac{\partial}{\partial c}\left(\frac{Z}{\rho}\right)\right\}^2 \tag{24-58}$$

The effecitve thermal conductivity κ_{eff} relates ∇T to the heat current Q

$$Q = -\kappa_{eff} \nabla T$$

By using Eqs. 24-21 and 24-22 we may calculate κ_{eff} for weak solutions

$$\kappa_{\text{eff}} = \kappa + \frac{\rho \mathfrak{D} m_3}{kc}\left\{\left(\sigma_0 + \frac{kc}{m_3}\right) - \frac{k}{m_3} k_T\right\}^2 \tag{24-60}$$

(σ_0 is the entropy per unit mass of pure helium II). For sufficiently weak solutions the second term in Eq. 24-60 is inversely proportional to the concentration and is larger than the thermal conductivity κ. Let us note that in pure helium, in the presence of ∇T there are no stationary solutions to the hydrodynamic equations. Under stationary conditions, the pressure p and potential μ are constant, and, therefore, the temperature is constant in helium.

Let us write down in their final form the hydrodynamic equations including dissipative terms, for solutions in helium II

$$\dot{\rho} + \operatorname{div} \mathbf{j} = 0$$

$$\frac{\partial j_i}{\partial t} + \frac{\partial \Pi_{ik}}{\partial r_k} = \frac{\partial}{\partial r_k}\left\{\eta\left(\frac{\partial v_n}{\partial r_k} + \frac{\partial v_{nk}}{\partial r_i} - \tfrac{2}{3}\delta_{ik}\frac{\partial v_{n\ell}}{\partial r_\ell}\right)\right\}$$

$$+ \frac{\partial}{\partial r_i}\{\zeta_1 \operatorname{div}(\mathbf{j} - \rho \mathbf{v}_n) + \zeta_2 \operatorname{div} \mathbf{v}_n\} \tag{24-61}$$

$$\frac{\partial}{\partial t}\rho c + \operatorname{div} \rho c \mathbf{v}_n = \operatorname{div}\left\{\rho \mathfrak{D}\left(\nabla c + \frac{k_T}{T}\nabla T + \frac{k_p}{p}\nabla p\right)\right\}$$

$$\dot{\mathbf{v}}_s + \nabla\left(\mu - \frac{Z}{\rho}c + \frac{v_s^2}{2}\right) = \nabla\{\zeta_3 \operatorname{div}(\mathbf{j} - \rho \mathbf{v}_n) + \zeta_4 \operatorname{div} \mathbf{v}_n\}$$

The equation of increase of entropy takes the form

$$T\left\{\dot{S} + \operatorname{div}\left(S\mathbf{v}_n + \frac{\mathbf{q} - \frac{Z\mathbf{g}}{\rho}}{T}\right)\right\} = R \tag{24-62}$$

$$R = \zeta_3[\operatorname{div}(\mathbf{j} - \rho \mathbf{v}_n)]^2 + \zeta_2[\operatorname{div} \mathbf{v}_n]^2 + 2\zeta_1 \operatorname{div}(\mathbf{j} - \rho \mathbf{v}_n)\operatorname{div} \mathbf{v}_n$$

$$+ \tfrac{1}{2}\eta\left(\frac{\partial v_{ni}}{\partial r_k} + \frac{\partial v_{nk}}{\partial r_i} - \tfrac{2}{3}\delta_{ik}\frac{\partial v_{n\ell}}{\partial r_\ell}\right)^2 + \frac{\kappa(\nabla T)^2}{T}$$

$$+ \rho \mathfrak{D}\frac{\partial}{\partial c}\left(\frac{Z}{\rho}\right)\left[\nabla c + \frac{k_T}{T}\nabla T + \frac{k_p}{p}\nabla p\right]^2 \tag{24-63}$$

SOUND IN SOLUTIONS IN HELIUM II

Let us use the system of equations 24-8 and 24-12 to find the velocity of sound in solutions of arbitrary concentrations. The linearized system may be written as

$$\frac{\partial \rho}{\partial t} + \operatorname{div} \mathbf{j} = 0$$

$$\frac{\partial \mathbf{j}}{\partial t} + \nabla p = 0$$

$$\frac{\partial \rho\sigma}{\partial t} + \rho c \operatorname{div} \mathbf{v}_n = 0 \qquad (24\text{-}65)$$

$$\frac{\partial \rho c}{\partial t} + \rho c \operatorname{div} \mathbf{v}_n = 0$$

$$\frac{\partial \mathbf{v}_s}{\partial t} + \nabla\left(\mu - \frac{Z}{\rho} c\right) = 0$$

If we eliminate the velocities $\mathbf{v}_n$ and $\mathbf{v}_s$ from Eqs. 24-65 we find three equations

$$\frac{\rho_n}{\rho_s} \frac{\ddot{\sigma}}{\sigma} = \sigma \, \Delta T + c \, \Delta \frac{Z}{\rho}$$

$$\ddot{\rho} = \Delta p \qquad (24\text{-}66)$$

$$\frac{\dot{c}}{c} = \frac{\dot{\sigma}}{\sigma}$$

Let us take as our independent variables the temperature T, the pressure p, and the concentration c. In a sound wave we may represent these quantities by their constant equilibrium value plus a small deviation, which we shall denote by the same letter with a prime. We look for a solution of Eq. 24-66 corresponding to a plane wave; in that case the deviations vary as $\exp[-i\omega(t - x/u)]$ (x is the direction of propagation of the wave, u the velocity of sound). If all the variables in Eq. 24-66 are taken of this form, we get

$$\left\{\frac{\rho_n}{\rho_s\sigma}\frac{\partial\sigma}{\partial T}u^2 - \sigma - c\frac{\partial}{\partial T}\frac{Z}{\rho}\right\}T' + \left\{\frac{\rho_n}{\rho_s\sigma}\frac{\partial\sigma}{\partial p}u^2 - c\frac{\partial}{\partial p}\frac{Z}{\rho}\right\}p'$$

$$+\left\{\frac{\rho_n}{\rho_s\sigma}\frac{\partial\sigma}{\partial c}u^2 - c\frac{\partial}{\partial c}\frac{Z}{\rho}\right\}c' = 0$$

$$\frac{\partial\rho}{\partial T}u^2T' + \left(\frac{\partial\rho}{\partial p}u^2 - 1\right)p' + \frac{\partial\rho}{\partial c}u^2c' = 0 \tag{24-67}$$

$$c\frac{\partial\sigma}{\partial T}T' + c\frac{\partial\sigma}{\partial p}p' + \left(c\frac{\partial\sigma}{\partial c} - \sigma\right)c' = 0$$

The compatibility condition for the system yields an equation which determines the velocity of sound in the solution

$$u^4\frac{\rho_n}{\rho_s}\left\{\frac{d\sigma}{dT}\frac{d\rho}{dp} - \frac{d\sigma}{dp}\frac{d\rho}{dT}\right\} - u^2\left\{\frac{d\rho}{dp}\left(\sigma + c\frac{\partial}{\partial T}\frac{Z}{\rho}\right)\sigma\right.$$

$$\left. + \frac{\rho_n}{\rho_s}\frac{d\sigma}{dT} - \frac{d\rho}{dT}\frac{d}{dp}\left(\frac{Z}{\rho}\right)\sigma c\right\} + \sigma\left(\sigma + c\frac{d}{dT}\frac{Z}{\rho}\right) = 0 \tag{24-68}$$

In order to condense the notation we introduce the total derivatives

$$\frac{df}{dT} = \frac{\partial f}{\partial T} + \frac{\partial f}{\partial c}\frac{c}{\overline{\sigma}}\frac{\partial\sigma}{\partial T} \qquad \frac{df}{dp} = \frac{\partial f}{\partial p} + \frac{\partial f}{\partial c}\frac{c}{\overline{\sigma}}\frac{\partial\sigma}{\partial p} \tag{24-69}$$

$$\overline{\sigma} = \sigma - c\frac{\partial\sigma}{\partial c} \tag{24-70}$$

(f is one of the three thermodynamic functions σ, ρ, Z/ρ). Equation 24-68 is too cumbersome to be easily examined. However we may use the simplifying fact that in practice the derivative $\partial\rho/\partial T$ is always extremely small, and neglect all terms in Eq. 24-68 which contain this derivative. Furthermore, let us use the relations which follow from the thermodynamic identity

$$d\mu = \frac{1}{\rho}dp - \sigma\, dT + \frac{Z}{\rho}dc$$

namely

$$\frac{\partial}{\partial p}\frac{Z}{\rho} = -\frac{1}{\rho^2}\frac{\partial \rho}{\partial c} \qquad \frac{\partial \sigma}{\partial p} = \frac{1}{\rho^2}\frac{\partial \rho}{\partial T} \qquad \frac{\partial \sigma}{\partial c} = -\frac{\partial}{\partial T}\frac{Z}{\rho} \tag{24-71}$$

As a result Eq. 24-68 takes the form

$$u^4 - u^2\left\{\frac{\rho_s}{\rho_n}\left[\frac{\bar{\sigma}^2}{\partial\sigma/\partial T} + c^2\frac{\partial}{\partial c}\left(\frac{Z}{\rho}\right)\right] + \left[1 + \frac{\rho_s}{\rho_n}\left(\frac{c}{\rho}\frac{\partial \rho}{\partial c}\right)^2\right]\bigg/\frac{\partial \rho}{\partial p}\right\}$$

$$+ \frac{\rho_s}{\rho_n}\left[\frac{\bar{\sigma}^2}{\partial\sigma/\partial T} + c^2\frac{\partial}{\partial c}\left(\frac{Z}{\rho}\right)\right]\bigg/\frac{\partial \rho}{\partial p} \tag{24-72}$$

Let us solve this equation for u^2 and use the fact that one of the roots u_1^2, which determines the velocity of first sound is much larger than the second u_2^2, which determines the velocity of second sound. As a result we have $(u_2^2 \ll u_1^2)$

$$u_1^2 = \left(\frac{\partial p}{\partial \rho}\right)_{c,T}\left[1 + \frac{\rho_s}{\rho_n}\left(\frac{\partial \rho}{\partial c}\frac{c}{\rho}\right)^2\right] \tag{24-73}$$

$$u_2^2 = \frac{\rho_s}{\rho_n}\left[\bar{\sigma}^2\left(\frac{\partial T}{\partial \sigma}\right)_{c,p} + c^2\frac{\partial}{\partial c}\left(\frac{Z}{\rho}\right)\right]\bigg/\left[1 + \frac{\rho_s}{\rho_n}\left(\frac{\partial \rho}{\partial c}\frac{c}{\rho}\right)^2\right] \tag{24-74}$$

To a first approximation in the concentration c the velocity of first sound is independent of the concentration.† The velocity of second sound contains terms which are linear in the concentration, namely $\bar{\sigma}$ and $c^2(\partial/\partial c)(Z/\rho)$, (the potential Z depends logarithmically on the concentration). The quantities $\bar{\sigma}$ and $(\partial/\partial c)(Z/\rho)$ which determine the velocity of second sound (Eq. 24-74) may be calculated for ideal solutions

$$\bar{\sigma} = \sigma - c\frac{\partial \sigma}{\partial c} = \sigma_{40} - \frac{k}{m_4}\left[c + \ln(1-c)\right] + \frac{k}{m_3}c \tag{24-75}$$

†We must emphasize that this result depends on the assumption that $\partial\rho/\partial T$ is small, since the exact expression for u_1^2 which follows from (24-68) contains terms linear in the concentration and proportional to $(\partial\rho/\partial T)(\partial\rho/\partial c)$.

$$c^2 \frac{\partial}{\partial c} \frac{Z}{\rho} = kT \left[\frac{c^2}{m_4 (1 - c)} + \frac{c}{m_3} \right] \tag{24-76}$$

For weak solutions, Eqs. 24-75 and 24-76 become

$$\overline{\sigma} = \sigma_{40} + \frac{kc}{m_3} \qquad c^2 \frac{\partial}{\partial c} \frac{Z}{\rho} = \frac{kTc}{m_3}$$

and Eq. 24-74 becomes†

$$u_2^2 = \frac{\rho_s}{\rho_n} \left[\frac{\partial T}{\partial \sigma} \left(\sigma_{40} + \frac{kc}{m_3} \right)^2 + \frac{kTc}{m_3} \right] \tag{24-77}$$

†The molar concentration $\epsilon = N_3/(N_3 + N_4)$ is related to the concentration c occuring in our formulas, by the formula $(1/\epsilon - 1) = (m_3/m_4)(1/c - 1)$.

25

THE DIFFUSION COEFFICIENT AND THERMAL CONDUCTIVITY IN WEAK SOLUTIONS OF HE^3 IN HELIUM II[31]

In order to calculate the kinetic coefficients of solutions it is necessary to find the distribution function in the solution, in the presence of small gradients of the thermodynamic functions and the velocities. Obviously, this problem can only be solved for weak solutions, when the impurity excitations may be treated as an ideal gas. We shall limit our considerations to the phenomena of diffusion and thermal conduction, which are intimately linked with one another, as can be seen from Eq. 24-60. Together they determine the heat transfer properties of solutions. The kinetic equation determining the distribution function f of the excitations in the solution, has the usual form of Eq. 18-1.

$$\frac{\partial n}{\partial t} + \frac{\partial n}{\partial \mathbf{r}}\cdot\frac{\partial H}{\partial \mathbf{p}} - \frac{\partial n}{\partial \mathbf{p}}\cdot\frac{\partial H}{\partial \mathbf{r}} = \mathcal{g}(n) \tag{25-1}$$

According to Eq. 18-2 the Hamiltonian of an excitation is equal to

$$H = \epsilon + \mathbf{p}\cdot\mathbf{v}_s \tag{25-2}$$

where ϵ is the energy of the excitation in the frame of reference in which $\mathbf{v}_s = 0$. In the presence of two modes of motion with velocities $\mathbf{v}_n$ and $\mathbf{v}_s$ the equilibrium distribution functions depend on the argument $(\epsilon - \mathbf{p}\cdot\mathbf{v}_n + \mathbf{p}\cdot\mathbf{v}_s)/kT$. The distribution functions of the phonons and rotons are the same as in Eqs. 1-13 and 1-14. The distribution function of the impurity excitations is equal to

$$n_i = \left(\frac{c\rho}{m_3}\right)(2\pi m^* kT)^{-3/2} \exp\left(-\frac{\epsilon + \mathbf{p}\cdot(\mathbf{v}_n - \mathbf{v}_s)}{kT}\right) \tag{25-3}$$

(all quantities referring to the impurities will carry the index i).

Let us suppose that a temperature gradient ∇T and a concentration gradient ∇c are present in the solution. The non-equilibrium distribution functions will then be determined by solving the kinetic equation. We shall assume, as is usually done, that at each point in space the system is in local equilibrium, i.e., that at each point the distribution function is to first order equal to the equilibrium function with local values of the temperature and concentration. For small values of the gradients ∇T and ∇c the changes in the other quantities will be determined by the linearized equations of hydrodynamics appropriate to solutions. If we insert the distribution function for the impurities (Eq. 25-3) into the left-hand side of the kinetic equation (Eq. 25-1), and express all time derivatives in terms of gradients by means of the hydrodynamic equations, we obtain the following kinetic equation

$$-\frac{n_i}{kT}\left\{\left(\frac{\rho}{\rho_n}\frac{kT}{m_3}p_i - \frac{kT}{c}\frac{\partial\epsilon_i}{\partial p_i}\right)\nabla T + \left[\frac{\rho}{\rho_n}\left(\sigma_0 + \frac{kc}{m_3}\right)p_i + \frac{\partial\epsilon_i}{\partial p_i}\right.\right.$$

$$\left.\left.\times\left(\frac{3}{2}k - \frac{\epsilon_i}{T}\right)\right]\nabla T\right\} = \mathcal{I}(n) \tag{25-4}$$

In a similar manner, the distribution functions of the phonons and rotons may be obtained

$$\frac{n'}{kT}\left\{\frac{\rho}{\rho_n}\frac{kT}{m_3}\mathbf{p}\cdot\nabla c + \left[\frac{\rho}{\rho_n}\left(\sigma_0 + \frac{kc}{m_3}\right)\mathbf{p} - \frac{\epsilon}{T}\frac{\partial\epsilon}{\partial\mathbf{p}}\right]\cdot\nabla T\right\} = \mathcal{I}(n) \tag{25-5}$$

Here n' is the derivative of the distribution function (of phonons or rotons) with respect to its argument. In order to solve the system of kinetic equations given in Eqs. 25-4 to 25-5 it is necessary to know the interactions between the elementary excitations (phonons, rotons, and impurities). The calculation of the corresponding differential and total cross-sections may be performed in the same way as was done in Chapter 7 for the case of roton-phonon interactions.

ROTON-IMPURITY SCATTERING

The interaction energy of an impurity and a roton can be expressed in the form of a δ-function of the distance between them

$$V = V_{01}\ \delta(r - r_1) \tag{25-6}$$

Using perturbation theory we find the following expression for the angular average of the effective cross-section

$$\sigma_{ir} = (|V_{01}|^2 m^*/2\pi\hbar^4)\, m^*\mu/(m^* + \mu) \tag{25-7}$$

Here m^* is the effective mass of the impurity μ the effective mass of the roton.

IMPURITY-IMPURITY SCATTERING

The interaction energy of two impurities may also be expressed as a δ-function, but with an amplitude V_{02} which naturally differs from V_{01}. Once again perturbation theory yields the cross-section

$$\sigma_{ii} = |V_{02}|^2 m^{*2}/4\pi\hbar^4 \tag{25-8}$$

PHONON-IMPURITY SCATTERING

Since the wavelength of the phonons is larger than the de Broglie wavelength of the impurities, the enternal structure of the impurities is not important. We may treat the impurity as a particle in the phonon field. The interaction energy for this problem is

$$V = -\tfrac{1}{2}(\mathbf{p}\cdot\mathbf{v} + \mathbf{v}\cdot\mathbf{p}) + \frac{\partial\epsilon_0}{\partial\rho}\rho' + \frac{1}{2}\frac{\partial^2\epsilon_0}{\partial\rho^2}\rho'^2 \tag{25-9}$$

where $\mathbf{p}$ is the momentum operator of the impurity, $\mathbf{v}$ the velocity of the medium, ρ' the deviation of the density of the solution from its equilibrium value, due to the presence of a phonon, and ϵ_0 the zero-point energy of an impurity excitation.

The differential cross-section for phonon-impurity scattering by an angle ψ, calculated with the interaction energy (Eq. 25-9), has the following form

$$d\sigma_{phi} = \left(\frac{pk^2}{4\pi\hbar^2\rho c}\right)^2 \left\{\frac{2}{3}(1 + \cos\psi)\cos^2\psi + \frac{p}{3\mu_c}(A + B)\right.$$

$$\left. \times (1 + \cos 2\psi) + (A + B)^2\right\} do_\psi \tag{25-10}$$

$$A = \frac{\rho^2}{pc}\left(\frac{\partial\epsilon_0}{\partial\rho}\right)\frac{1}{\mu c^2}\cos\psi + \frac{\rho^2}{pc}\left(\frac{\partial\epsilon_0}{\partial\rho}\right)\frac{1}{\rho^2}$$

$$\times\left[(1 - \cos\psi)\frac{c^2}{\rho} - \frac{\partial c^2}{\partial\rho}\right] \qquad B = \frac{\rho^2}{pc}\frac{\partial^2\epsilon_0}{\partial\rho^2}$$

k is the phonon momentum and c the velocity of sound.

DIFFUSION

We shall calculate the diffusion coefficient in a number of limiting cases in which it is possible to solve the kinetic equation. First, we shall consider the region of relatively high temperatures, where in pure helium the rotons are dominant. In this temperature region it is possible to solve the kinetic equation for very small concentrations, when the number of impurities is much less than the number of rotons, and also in the opposite limit, when the number of impurities is much greater than the number of rotons. Then we consider relatively low temperatures, at which there are few rotons, and only phonon-impurity interactions are important. In this temperature region the problem may also be solved in two limiting cases, when the number of impurities is much larger than the number of phonons, and when it is much smaller.

We shall not perform here the rather cumbersome calculations, but merely write down the final result for the diffusion coefficient

$$\mathfrak{D} = \left(\frac{\rho_{n0}}{\rho_n}\right)^2 \frac{kT}{m_3} t_i \tag{25-11}$$

which is actually an interpolation between the results obtained in the aforementioned limiting cases. Here ρ_n is the total normal density, ρ_{n0} the contribution of phonons and rotons to the normal density, and t_i the average collision time of impurities with phonons and rotons; it is equal to

$$t_i = \left\{\overline{\sigma_{ir} v_i}\, \mathfrak{N}_r + \overline{\sigma_{phi} c}\, \mathfrak{N}_{ph} \frac{4kT}{m^* c^2} \frac{1 + 8t_{phi}/\theta_{ph}}{10 + 8t_{phi}/\theta_{ph}}\right\} \tag{25-12}$$

The average phonon-impurity collision time t_{phi} is given by

$$t_{phi} \approx 2/\overline{\sigma_{phi} c}\, \mathfrak{N}_i \tag{25-13}$$

Finally, the time θ_{ph} characterizes the establishment of equilibrium in the number of phonons. It was introduced by us earlier, in the discussion of kinetic coefficient in pure helium (cf. Eq. 19-16).

The meaning of Eq. 25-11 is quite clear, since apart from the dimensionless factor $[(\rho_{n0})/\rho]^2$, it may be obtained by simple kinetic theory arguments, according to which we have

$$\mathfrak{D} \sim v_i \ell_i \sim v_i^2 t_i \sim \frac{kT}{m_3} t_i$$

THERMAL DIFFUSION

The thermal diffusivity k_T only enters into the expression for the impurity current

$$g = -\rho \mathfrak{D} \left(\nabla c + \frac{k_T}{T} \nabla T \right) \tag{25-14}$$

for relatively high concentrations, when the number of impurity excitations is larger than the number of phonons or rotons. This phenomenon may be studied by the same method as diffusion; the result is

$$k_T = c \left(1 - \frac{\rho_{ni}}{\rho_{n0}} \frac{\sigma_0 m_3}{kc} \right) \tag{25-15}$$

THERMAL CONDUCTION

In solutions, just as in pure helium, a temperature gradient causes an additional heat flux whose magnitude is determined by the thermal conductivity κ. In solutions, this coefficient may be divied into three parts, referring to phonons, rotons and impurities, respectively. These may be calculated in the same way as the corresponding quantities in pure helium II. We shall not write down the complicated formulas one obtains, but merely note that here also the main contribution comes from the phonon part, which varies as $\epsilon^{\Delta/kT}$ over a wide temperature range. The rapid increase of κ_{ph} with decreasing temperature is due to the increase in the phonon mean free path.

THE EFFECTIVE THERMAL CONDUCTIVITY OF SOLUTIONS

In a solution, for a given temperature gradient, there automatically arises a concentration gradient. Therefore, as we saw in Chapter 24, the heat flux which arises under stationary conditions in solutions, is determined by some combination of the diffusion constant, the thermal diffusivity and the thermal conductivity. This combination, which we called the effective thermal conductivity (κ_{eff}), has been calculated in Eq. 24-60. Inserting the expressions for $\mathfrak{D}$ and k_T given in Eqs. 25-11 and 25-14 into Eq. 24-60 we obtain the effective thermal conductivity for dilute solutions

$$\kappa_{eff} = \rho T \sigma_0^2 t_i / c + \kappa \tag{25-16}$$

(σ_0 is the entropy of the system apart from the impurity contribution).

The first term in Eq. 25-16 is due to diffusion and thermal diffusion. The average collision time of impurities with phonons and rotons depends weakly on the concentration, so that the first term in κ_{eff} increases as $1/c$ with decreasing concentration. The first term decreases rapidly with decreasing temperature, since the entropy decreases. On the other hand the second term (κ), increases with decreasing temperature, as we already mentioned, so that the effective thermal conductivity (κ_{eff}) has a characteristic minimum which is clearly visible in the figure given in Ref. 31.

REFERENCES

1. L. Landau, J. Phys. USSR **5**, 71 (1941). The theory of superfluidity of helium II. Reprinted in this volume.
2. L. Landau, J. Phys. USSR **11**, 91 (1947). On the theory of superfluidity of helium II. Reprinted in this volume.
3. L. Landau, Phys. Rev. **75**, 884 (1949). On the theory of superfluidity.
4. D. Henshaw and A. Woods, Phys. Rev. **121**, 1266 (1961). Modes of atomic motions in liquid helium II.
   H. Palevsky, K. Otnes, and K. E. Larsson, Phys. Rev. **112**, 11 (1958). Excitation of rotons in helium II by cold neutrons.
   J. L. Yarnell, G. P. Arnold, P. J. Bendt, and E. C. Kerr, Phys. Rev. **113**, 1379 (1959). Excitations in liquid helium: neutron scattering measurements.
   P. J. Bendt, R. D. Cowan, and J. L. Yarnell, Phys. Rev. **113**, 1386 (1959). Excitations in liquid helium: thermodynamic calculations.
5. R. P. Feynman, Phys. Rev. **94**, 262 (1954). Atomic theory of the 2-fluid model of helium II.
   A. Bijl, Physica **7**, 869 (1946).
6. L. P. Pitaevskii, Zh. Eksperim. i Teor. Fiz. **31**, 536 (1956). [Translation: Soviet Phys. JETP **4**, 439 (1956).] On the derivation of the energy spectrum of liquid helium II.
7. N. N. Bogoliubov, J. Phys. USSR **11**, 23 (1947). On the theory of superfluidity. Reprinted in D. Pines, "The Many-Body Problem," W. A. Benjamin, New York (1961).
8. T. D. Lee and C. N. Yang, Phys. Rev. **105**, 1119 (1957). Many-body problem in quantum mechanics and quantum statistical mechanics.
9. L. P. Pitaevskii, Zh. Eksperim. i Teor. Fiz. **36**, 1168 (1959). [Translation: Soviet Phys. JETP **9**, 830 (1959).] On the properties of the elementary excitation spectrum in the vicinity of the decay threshold of the excitations.

10. L. D. Landau and I. M. Khalatnikov, Zh. Eksperim. i Teor. Fiz. **19**, 637 (1949). The theory of the viscosity of helium II: I) Collisions of elementary excitations. (In Russian.)
11. I. M. Khalatnikov, Zh. Eksperim. i Teor. Fiz. **20**, 243 (1950). The absorption of sound in helium II. (In Russian.)
12. I. M. Khalatnikov, Zh. Eksperim. i Teor. Fiz. **23**, 169 (1952). The hydrodynamics of solutions of impurities in helium II. (In Russian.)
13. I. M. Khalatnikov, Zh. Eksperim. i Teor. Fiz. **23**, 8 (1952). Kinetic coefficients in helium II. (In Russian.)
14. E. M. Lifshitz, J. Phys. USSR **8**, 110 (1944). Radiation of sound in helium II.
15. V. P. Peshkov, Zh. Eksperim. i Teor. Fiz. **18**, 857 (1948). The excitation and propagation of second sound. (In Russian.)
16. I. M. Khalatnikov, Zh. Eksperim. i Teor. Fiz. **23**, 21 (1952). Thermal conductivity and absorption of sound in helium II. (In Russian.)
17. I. M. Khalatnikov, Zh. Eksperim. i Teor. Fiz. **23**, 253 (1952). Discontinuities and large amplitude sound waves in helium II. (In Russian.)
18. K. R. Atkins, Phys. Rev. **113**, 962 (1959). Third and fourth sound in liquid helium II.
19. I. Bekarevitch and I. M. Khalatnikov, Zh. Eksperim. i Teor. Fiz. **40**, 920 (1961). [Translation: Soviet Phys. JETP **13**, 643 (1961).] Phenomenological derivation of the equations of vortex motion in helium II.
20. R. P. Feynman, in "Progress in Low Temperature Physics" (C. J. Gorter, ed.), Vol. I, Chap. II, North Holland, Amsterdam (1955). Application of quantum mechanics to liquid helium.
21. H. Hall, Phil. Mag. Suppl. **9**, No. 33, 89 (1960). W. F. Vinen, in "Progress in Low Temperature Physics" (C. J. Gorter, ed.), Vol. III, Chap. I, North Holland, Amsterdam (1961). Vortex lines in liquid helium II.
22. L. P. Pitaevskii, Zh. Eksperim. i Teor. Fiz. **35**, 408 (1958). [Translation: Soviet Phys. JETP **8**, 282 (1959).] Phenomenological theory of superfluidity in the vicinity of the λ-point.
23. L. D. Landau and I. M. Khalatnikov, Dokl. Akad. Nauk SSR **96**, No. 3, 469 (1954). On the anomalous absorption of sound in the vicinity of a second order phase transition.
24. L. D. Landau and I. M. Khalatnikov, Zh. Eksperim. i Teor. Fiz. **19**, 709 (1949). The theory of the viscosity of helium II: II) The calculation of the viscosity coefficient. (In Russian.)
25. A. Andreyev and I. M. Khalatnikov, Zh. Eksperim. i Teor. Fiz. **44**, 2058 (1963). [Translation: Soviet Phys. JETP **17**, 1384 (1963).] Sound in liquid helium II near zero temperature.
26. I. M. Khalatnikov, Zh. Eksperim. i Teor. Fiz. **22**, 687 (1952). Heat exchange between a solid and helium II. (In Russian.)

27. L. D. Landau and I. Pomeranchuk, Dokl. Akad. Nauk SSR **59**, 668 (1948). On the motion of impurities in helium II. (In Russian.)
28. I. Pomeranchuk, Zh. Eksperim. i Teor. Fiz. **19**, 42 (1949). The influence of impurities on the thermodynamic properties and the velocity of second sound in helium II. (In Russian.)
29. I. Khalatnikov, Zh. Eksperim. i Teor Fiz. **23**, 265 (1952). Sound in solutions of impurities in helium II and dissipative functions of solutions. (In Russian.)
30. R. Arkhipov and I. M. Khalatnikov, Zh. Eksperim. i Teor. Fiz. **33**, 758 (1957). [Translation: Soviet Phys. JETP **6**, 583 (1958).] Propagation of sound across a boundary between two superfluid phases.
31. I. M. Khalatnikov and V. Zharkov, Zh. Eksperim. i Teor. Fiz. **32**, 1108 (1957). [Translation: Soviet Phys. JETP **5**, 905 (1957).] Theory of diffusion and thermal conductivity of dilute solutions in helium II.

SUPPLEMENTARY REFERENCES

An extensive bibliography appears in K. R. Atkins, "Liquid Helium" (Cambridge Univ. Press, Cambridge, 1959).

Since the Soviet literature is not fully represented in Atkins we reprint here the Liquid Helium section of the general index of the Zhurnal Eksperimental'noi i Teoreticheskoi Fiziki for 1946–1961 [Zh. Eksperim. i Teor. Fiz. **45**, No. 1 (1963); translation: Soviet Phys. JETP **18**, 1474 (1964)], as well as the appropriate references to more recent Soviet work.

Some of the articles, whose translations do not appear in Soviet Physics JETP, have been translated by other sponsors. These are denoted as follows:

ATS	Associated Technical Services, Inc. P.O. Box 271 East Orange, New Jersey 088
LC	Photoduplication Service Publication Board Project Library of Congress Washington, D.C. 20025
SLA	Special Libraries Association Translation Center 35 W. 33rd St. Chicago, Illinois 60616

HE^4 (EXPERIMENT)

Determination of the velocity of propagation of second sound in helium II. V. P. Peshkov—[**16**, 1000 (1946)].

Direct observation of two kinds of motion in helium II. É. Andronikashvilli [**16**, 780 (1946)].

Scattering of low-energy neutrons in helium II. A. Akhiezer and I. Pomeranchuk [**16**, 391 (1946)].

The temperature dependence of the normal density of helium II. É. L. Andronikashvili [**18**, 424 (1948)]. LC, SLA
Investigation of the viscosity of the normal component of helium II. E. L. Andronikashvili [**18**, 429 (1948)]. LC, SLA, ATS
Second sound in helium II at elevated pressures. V. P. Peshkov and K. N. Zinov'eva [**18**, 438 (1948)]. SLA
The achievement of temperatures below 1°K by vapor cooling under liquid helium. B. N. Esel'son [**18**, 795 (1948)]. SLA
Conditions for the excitation and propagation of second sound. V. P. Peshkov [**18**, 857 (1948)].
Velocity of second sound from 1.3 to 1.03°K. V. P. Peshkov [**18**, 951L (1948)].
A study of the properties of second sound. V. P. Peshkov [**18**, 867 (1948)]. SLA
Dispersion in second sound at low frequencies. V. P. Peshkov [**19**, 270 (1949)].
Concerning heat transfer in helium II. E. L. Andronikashvili [**19**, 533 (1949)]. LC, SLA
Some problems of the hydrodynamics of helium II. E. L. Andronikashvili [**22**, 62 (1952)]. LC, SLA
New features of the transport effect on a film of He II. B. N. Esel'son and B. G. Lazarev [**23**, 552 (1952)]. SLA
The determination of the velocity of second sound in liquid helium II down to a temperature of 0.85°K. V. P. Peshkov [**23**, 686 (1952)].
Absorption of second sound in helium II. K. N. Zinov'eva [**25**, 235 (1953)].
Measurement of the thermomagnetic effect in helium II in the vicinity of 1°K. V. P. Peshkov [**27**, 351 (1954)]. SLA
Rotation of helium II at high speeds. E. L. Andronikashvili and I. P. Kaverkin—**1**, 174 (1955) [**28**, 126 (1955)].
The behavior of helium II in the neighborhood of a heat radiating surface. E. I. Andronikashvili and G. G. Mirskaya—**2**, 406 (1956) [**29**, 490 (1955)].
Transformation of the λ-transition in helium to a special transition of the first kind in the presence of a heat flow. V. P. Peshkov—**3**, 628 (1956) [**30**, 581 (1956)].
The coefficient of volume absorption of second sound and the viscosity of the normal component of helium II down to 0.83°K. K. N. Zinov'eva—**4**, 36 (1956) [**31**, 31 (1956)].
Some properties of rotating He II. B. N. Esel'son, B. G. Lazarev, K. D. Sinel'nikov, and A. D. Shvets—**4**, 774 (1957) [**31**, 912 (1956)].
On the existence of a tangential velocity discontinuity in the superfluid component of helium near a wall. G. A. Gamtsemlidze—**7**, 992 (1958) [**34**, 1434 (1958)].
A method for the observation of helium II films. A. I. Shal'nikov—**8**, 386 (1959) [**35**, 558 (1958)].

Observation of the λ-transitions in helium in the presence of a thermal current through the phase boundary. V. P. Peshkov—**8**, 943 (1959) [**35**, 1350 (1958)].

The propagation of oscillations along vortex lines in rotating helium II. É. L. Andronkiashvili and D. S. Tsakadze—**10**, 227 (1960) [**37**, 322 (1959)].

The use of superconducting ring for registering the phase transition in liquid helium. B. N. Esel'son and A. D. Shvets—**10**, 228 (1960) [**37**, 323 (1959)].

Experimental investigation of the harmonic oscillations of a disk in rotating helium II. E. L. Andronkiashvili and D. S. Tsakadze—**10**, 397 (1960) [**37**, 562 (1959)].

On the critical mode in experiments with an oscillating disk in helium II. G. A. Gamtsemlidze—**10**, 678 (1960) [**37**, 950 (1959)].

On the problem of the motion of charges in liquid helium II. R. G. Arkhipov and A. I. Shal'nikov—**10**, 888 (1960) [**37**, 1247 (1959)].

Measurement of the logarithmic damping decrement of a hollow cylinder in rotating helium II. D. S. Tsakadze and I. M. Chkheidze—**11**, 457 (1960) [**38**, 637 (1960)].

Second sound in helium II. V. P. Peshkov—**11**, 580 (1960) [**38**, 799 (1960)].

HE⁴ (THEORY)

Theory of viscosity of helium II. II. Calculation of the viscosity coefficient. L. D. Landau and I. M. Khalatnikov [**19**, 709 (1949)]. SLA

Theory of viscosity of helium II. 1. Collision of elementary excitations in helium II. L. D. Landau and I. M. Khalatnikov [**19**, 637 (1949)]. LC, SLA

Absorption of sound in helium II. I. M. Khalatnikov [**20**, 243 (1950)]. LC, SLA

Concerning the dependence of the thickness of a He II film on the temperature. M. I. Kaganov and B. N. Esel'son [**21**, 656 (1951)]. LC, SLA

Heat exchange between solids in helium II. I. M. Khalatnikov [**22**, 687 (1952)]. SLA

Kinetic coefficients in helium II. I. M. Khalatnikov [**23**, 8 (1952)]. SLA, DSIR-HQ

Thermal conductivity and sound absorption in helium II. I. M. Khalatnikov [**23**, 21 (1952)]. SLA

Discontinuities and high amplitude sound in helium II. I. M. Khalatnikov [**23**, 253 (1952)]. SLA

The surface energy associated with a tangential velocity discontinuity in helium II. V. L. Ginzburg—**2**, 170 (1956) [**29**, 244 (1955)].

The effective density of rotating liquid helium II. I. M. Lifshitz and M. I. Kaganov—**2**, 172 (1956) [**29**, 257 (1955)].

The propagation of sound in moving helium II and the effect of a thermal current upon the propagation of second sound. I. M. Khalatnikov—**3**, 649 (1956) [**30**, 617 (1956)].

Effect of the rate of flow of a He II film on its thickness. V. M. Kontorovich—**3**, 770 (1956) [**30**, 805 (1956)].

On the derivation of a formula for the energy spectrum of liquid He^4. L. P. Pitaevskii—**4**, 439 (1957) [**31**, 536 (1956)].

Flow instability of a superfluid film. R. G. Arkhipov—**6**, 90 (1958) [**33**, 116 (1957)].

Behavior of particles of small effective mass in superfluid helium. R. G. Arkipov—**6**, 307 (1958) [**33**, 397 (1957)].

Absorption of second sound in rotating helium II. E. M. Lifshitz and L. P. Pitaevskii—**6**, 418 (1958) [**33**, 535 (1957)].

Hydrodynamic fluctuations in a superfluid liquid. I. M. Khalatnikov—**6**, 624 (1958) [**33**, 809 (1957)].

Thermodynamic functions of superfluid helium films. R. G. Arkhipov—**6**, 634 (1958) [**33**, 822 (1957)].

On the theory of superfluidity. V. L. Ginsburg and L. P. Pitaevskii—**7**, 858 (1958) [**34**, 1240 (1958)].

Phenomenological theory of superfluidity near the λ point. L. P. Pitaevskii—**8**, 282 (1959) [**35**, 408 (1958)].

Calculation of the phonon part of the mutual friction force in superfluid helium. L. P. Pitaevskii—**8**, 888 (1959) [**35**, 1271 (1958)].

Properties of the spectrum of elementary excitations near the disintegration threshold of the excitations. L. P. Pitaevskii—**9**, 830 (1959) [**36**, 1168 (1959)].

On the question of critical velocities for flow of He II in capillaries. B. T. Geĭlikman—**10**, 635 (1960) [**37**, 891 (1959)].

Damping of oscillations of a disc in rotating helium II. Yu. G. Mamaladze and S. G. Matinyan—**11**, 134 (1960) [**38**, 184 (1960)].

Damping of oscillations of a cylinder in rotating helium II. Yu. G. Mamaladze and S. G. Matinyan—**11**, 471 (1960) [**38**, 656 (1960)].

The problem of the form of the spectrum of elementary excitations of liquid helium II. L. P. Pitaevskiĭ—**12**, 155 (1961) [**39**, 216 (1960)].

On mutual friction in helium II. Yu. G. Mamaladze—**12**, 595 (1961) [**39**, 859 (1960)].

HE^3 AND HE^3-HE^4 MIXTURES (EXPERIMENT)

Some properties of solutions of He^3 in He^4. I. Separation of helium isotopes. B. N. Esel'son and B. G. Lazarev [**20**, 742 (1950)].

Some properties of solutions of He^3 in He^4. II. Shift of λ point and features of the transport effect. B. N. Esel'son, B. G. Lazarev, and I. M. Lifshitz [**20**, 748 (1950)].

Measurement of vapor tension over solutions of He^3 in He^4. B. N. Esel'son, B. G. Lazarev, and N. E. Alekseevskiĭ [**20**, 1055 (1950)]. LC, SLA

Some properties of solutions of He^3 in He^4. III. The vapor pressure. B. N. Esel'son [**26**, 744 (1954)]. SLA

Dew point of mixture of helium isotopes. B. N. Esel'son and N. G. Berznyak [**27**, 648 (1954)]. ATS

The surface tension of liquid He^3 in the temperature range 0.93 - 3.34°K. K. N. Zinov'eva—**1**, 173 (1955) [**28**, 125 (1955)].

The surface tension of liquid He^3 in the region of very low temperatures (1.0—0.35°K). K. N. Zinov'eva—**2**, 774 (1956) [**29**, 899 (1955)].

Phase diagram for the liquid-vapor system of helium isotopes (He^3-He^4). I. B. N. Esel'son and N. G. Berezniak—**3**, 568 (1956) [**30**, 628 (1956)].

Experiments in enrichment of helium with isotope He^3. V. P. Peshkov —**3**, 706 (1956) [**30**, 850 (1956)].

Measurement of the saturated vapor pressure of a He^3-He^4 mixture with a high He^3 concentration. V. P. Peshkov and V. N. Kachinskii —**4**, 607 (1957) [**31**, 720 (1956)].

Density of the normal component for solutions of the isotopes of helium. N. G. Berezniak and B. N. Esel'son—**4**, 766 (1957) [**31**, 902 (1956)].

Optical determination of the density of He^3. V. P. Peshkov—**6**, 645 (1958) [**33**, 833 (1957)].

Density of He^3-He^4 solutions. T. P. Ptukha—**7**, 22 (1958) [**34**, 33 (1958)].

Film transfer rate in helium isotope mixtures. B. N. Esel'son, A. D. Shvets, and R. A. Bablidze—**7**, 161 (1958) [**34**, 233 (1958)].

Viscosity of liquid He^3 in the range 0.35–3.2°K and He^4 above the lambda-point. K. N. Zinov'eva—**7**, 421 (1958) [**34**, 609 (1958)].

He^3 cryostats. V. P. Peshkov, K. N. Zinov'eva, and A. I. Filimonov —**9**, 734 (1959) [**36**, 1034 (1959)].

Phase diagram for liquid He^3-He^4 solutions. K. N. Zinov'eva and V. P. Peshkov—**10**, 22 (1960) [**37**, 33 (1959)].

The determination of the coefficients of diffusion and of heat conductivity of weak solution of He^3 in helium II. T. P. Ptukha—**12**, 621 (1961) [**39**, 896 (1960)].

HE^3 AND HE^3-HE^4 MIXTURES (THEORY)

Effect of impurities on the thermodynamic properties and velocity of second sound in He II. I. Pomeranchuk [**19**, 42 (1949)]. LC, SLA

Contributions to the theory of liquid He^3. I. Pomeranchuk [**20**, 919 (1950)]. SLA

Specific heat of liquid He^3. E. M. Lifshitz [**21**, 659 (1951)]. LC, SLA

The hydrodynamics of solutions of foreign particles in helium II. I. M. Khalatnikov [**23**, 169 (1952)]. SLA

Sound in solutions of foreign particles in helium II and the dissipation function of solutions. I. M. Khalatnikov [**23**, 265 (1952)].

Hydrodynamics of solutions of two superfluid liquids. I. M. Khalatnikov—**5**, 542 (1957) [**32**, 653 (1957)].

The thermodynamics of liquid He^3. I. M. Khalatnikov and A. A. Abrikosov—**5**, 745 (1957) [**32**, 915 (1957)].

Theory of kinetic phenomena in liquid He^3. A. A. Abrikosov and I. M. Khalatnikov—**5**, 887 (1957) [**32**, 1083 (1957)].

Theory of diffusion and thermal conductivity for dilute solutions of He^3 in helium II. I. M. Khalatnikov and V. N. Zharkov—**5**, 905 (1957) [**32**, 1108 (1957)].

Dispersion of sound in a Fermi liquid. I. M. Khalatnikov and A. A. Abrikosov—**6**, 84 (1958) [**33**, 110 (1957)].

On the scattering of light in He^3-He^4 mixtures. L. P. Gor'kov and L. P. Pitaevskii—**6**, 486 (1958) [**33**, 634 (1957)].

Propagation of sound across a boundary between two superfluid phases. R. G. Arkhipov and I. M. Khalatnikov—**6**, 583 (1958) [**33**, 758 (1957)].

The influence of a He^3 impurity on the viscosity of helium II. V. N. Zharkov—**6**, 714 (1958) [**33**, 929 (1957)].

Thermodynamics of the He I—He II phase transitions in helium isotope mixtures. B. N. Esel'son, M. I. Kaganov, and I. M. Lifshitz—**6**, 719 (1958) [**33**, 936 (1957)].

Scattering of light in a Fermi liquid. A. A. Abrikosov and I. M. Khalatnikov—**7**, 135 (1958) [**34**, 198 (1958)].

Contributions to the theory of the Pomeranchuk effect in He^3. D. G. Sanikidze—**8**, 192 (1959) [**35**, 279 (1958)].

Pomeranchuk effect and diagram of state for He^3-He^4 solutions. I. M. Lifshitz and D. G. Sanikidze—**8**, 713 (1959) [**35**, 1020 (1958)].

Theory of weak solutions of He^4 in liquid He^3. V. N. Zharkov and V. P. Silin—**10**, 102 (1960) [**37**, 143 (1959)].

Hydrodynamics of solutions of strange particles in helium II near the λ-point. D. G. Sanikidze—**10**, 226 (1960) [**37**, 320 (1959)].

On the superfluidity of liquid He^3. L. P. Pitaevskii—**10**, 1267 (1960) [**37**, 1794 (1959)].

Theory of the Kapitza thermal discontinuity of the boundary between liquid He^3 and a solid body. I. L. Bekarevich and I. M. Khalatnikov—**12**, 1187 (1961) [**39**, 1699 (1960)].

The references given below are to Soviet Physics JETP.

Volume 13, July-Dec. 1961

HELIUM, LIQUID

Critical velocities in superfluid helium. V. P. Peshkov—259L.
Phenomenological derivation of the equations of vortex motion in He II. I. L. Bekarevich and I. M. Khalatnikov—643.
Thermal conductivity and diffusion of weak He^3-He^4 solutions in the temperature range from the λ-point to 0.6°K. T. P. Ptukha—1112.
Theory of absorption of sound in dilute He^3 in He II solutions. A. F. Andreev—1201.
Vortex lines in an imperfect Bose gas. L. P. Pitaevskii—451.

Volume 14, Jan.-June 1962

Kinetics of the Destruction of Superfluidity in Helium. V. P. Peshkov and V. K. Tkachenko—1019.
Motion of Charges in Liquid and Solid Helium. A. I. Shal'nikov—755.
On the Theory of the Scattering of Slow Neutrons in a Fermi Liquid. A. I. Akhiezer, I. A. Akhiezer, and I. Ya. Pomeranchuk—343.
Scattering of Gamma Rays in Liquid He^3. A. A. Abrikosov and I. M. Khalatnikov—389.
Some Observations on the Solidification of Helium. A. I. Shal'nikov—753.
Which is Responsible for the Destruction of Superfluidity, v_s or $v_s - v_n$? V. P. Peshkov and V. B. Stryukov—1031.

Volume 15, July-Dec. 1962

An investigation of the temperature discontinuity at the boundary between a solid and superfluid helium. Kuang Wey Yen—635.
Determination of the coefficient of mutual friction between the superfluid and normal components along vortex lines. D. S. Tsakadze—681.
λ-Points of concentrated He^3-He^4 mixtures. B. N. Esel'son, V. G. Ivantsov, and A. D. Shvets—651.
Singularities in the equilibrium diagram of a He^3-He^4 solution at the λ-point. D. G. Sanikidze—922.
The transition of liquid He^3 into the superfluid state. L. P. Gor'kov and L. P. Pitaevskiĭ—417.

Volume 16, Jan.-June 1963

An investigation of rotating He II near the λ-point using second sound. É. L. Andronkiashvili, D. S. Tsakadze, and R. A. Bablidze—1103L.

Curves of the commencement of solidification of helium isotope solutions. N. G. Bereznyak, I. V. Bogoyavlenskiĭ, and B. N. Esel'son—1394L.
The Supercritical oscillatory mode in He II. R. A. Bablidze—1476.

Volume 17, July-Dec. 1963

Effective mutual friction between the superfluid and normal components of He II along its axis of rotation. D. S. Tsakadze—**17**, 70 (1963) [**44**, 103 (1963)].
Formation of a central vortex in rotating liquid helium. D. S. Tsakadze—**17**, 72 (1963) [**44**, 105 (1963)].
Surface tension of He^3-He^4 solutions. B. N. Esel'son, V. G. Ivantsov, and A. D. Shvets—**17**, 330 (1963) [**44**, 483 (1963)].
Absorption of sound in liquid He II below 0.6°K. I. M. Khalatnikov—**17**, 519L (1963) [**44**, 769L (1963)].
Diagram of state for He^3-He^4 solutions: phase stratification and solidification curves. K. N. Zinov'eva—**17**, 1235 (1963) [**44**, 1837 (1963)].
Sound in liquid helium II near absolute zero. A. Andreev and I. Khalatnikov—**17**, 1384 (1963) [**44**, 2058 (1963)].
Stability of rotation of a superfluid liquid. Yu. G. Mamaladze and S. G. Matinyan—**17**, 1424 (1963) [**44**, 2118 (1963)].
Direct measurements of the linear flow rate of a He II film. B. N. Esel'son, Yu. Z. Kovdrya, and B. G. Lazarev—**17**, 1469L (1963) [**44**, 2187L (1963)].

Volume 18, Jan.-June 1964

The equilibrium diagram for the He^3-He^4 liquid-crystal system. (E) N. G. Bereznyak, I. V. Bogoyavlenskii, and B. N. Esel'son—**18**, 335 (1964) [**45**, 486 (1963)].
Development of turbulence in the presence of a heat flow in helium II within a capillary and the critical velocity problem. (E) V. K. Tkachenko—**18**, 1251 (1964) [**45**, 1827 (1963)].

Volume 19, July-Dec. 1964

Some Properties of the Central Macroscopic Vortex in Rotating Helium. (E) D. S. Tsakadze—**19**, 110 (1964) [**46**, 153 (1963)].
Possible Existence of Onsager-Feynman Vortices at Temperatures Above the λ Point. (E) E. L. Andronikashvili, K. B. Mesoed, and Dzh. S. Tsakadze—**19**, 113 (1964) [**46**, 157 (1963)].
Thermomechanical Effect Near the λ Point in Rotating Liquid

Helium. (E) L. A. Zamtaradze and Dzh. S. Tsakadze—**19**, 116 (1964) [**46**, 162 (1963)].

On the Irrotational Region in Rotating He II. (T/E) M. P. Kemoklidze and Yu. G. Mamaladze—**19**, 118 (1964) [**46**, 165 (1963)].

Attempt to Detect an Irrotational Region in Rotating He II. (E) D. S. Tsakadze—**19**, 343 (1964) [**46**, 505 (1963)].

The Shape of the Meniscus of Rotating He II. (T) M. P. Kemoklidze and Yu. G. Mamaladze—**19**, 547 (1964) [**46**, 804 (1963)].

On the Problem of the λ-Point Shift in Rotating Liquid Helium. (E) R. A. Bablidze, Dzh. S. Tsakadze, and G. V. Chanishvili—**19**, 577 (1964) [**46**, 843 (1963)].

Fourth Sound in an He^3-He^4 Solution. (T) D. G. Sanikidze and D. M. Chernikova—**19**, 760 (1964) [**46**, 1123 (1963)].

Properties of Rotating Liquid He in the Vicinity of the Lambda Point. (T) A. F. Andreev—**19**, 983 (1964) [**46**, 1456 (1964)].

Superfluidity of He^3. (E) V. P. Peshkov—**19**, 1023L (1964) [**46**, 1510L (1964)].

Kinetics of Growth of the Meniscus of a Rotating Liquid. (E) Dzh. S. Tsakadze—**19**, 1050 (1964) [**46**, 1553 (1964)].

Hydrodynamics of Rotating Helium II in an Annular Channel. (T) M. P. Kemoklidze and I. M. Khalatnikov—**19**, 1134 (1964) [**46**, 1677 (1964)].

Thermal Conductivity of Solid He^4. (E) L. P. Mezhov-Deglin—**19**, 1297L (1964) [**46**, 1926L (1964)].

PART V

REPRINTS

THE THEORY OF SUPERFLUIDITY OF HELIUM II

By L. LANDAU

(Received May 21, 1941)

The quantization of an arbitrary system of interacting particles (a liquid) is performed by means of introducing the operators of the density and of the velocity of the liquid; the commutation rules between these operators are determined (§ 1). From the results of this quantization the general character of the distribution of the energy levels in the spectrum of a quantum liquid is determined (§ 2). The temperature dependence of the heat capacity of helium II is investigated (§ 3). It is shown that at absolute zero a quantum liquid can possess the property of superfluidity (§ 4). At non-zero temperatures it is found that two motions — a superfluid and a normal — can simultaneously exist in helium II. This can be described by means of the conception of the superfluid and normal parts of the liquid; the λ-point in helium II is connected with the disappearance of the «superfluid» part of the liquid (§ 5). The experiments made to measure the heat conductivity and viscosity of helium II are interpreted; the thermomechanical effects in helium II are considered (§ 6). A system of hydrodynamic equations is advanced describing the macroscopic motion of helium II (§ 7). By means of these equations the propagation of sound is investigated and it is shown that two velocities of sound must exist in helium II (§ 8).

Liquid helium is known to possess a number of peculiar properties at temperatures lower than the λ-point. Of these properties the most important one is superfluidity discovered by P. L. Kapitza (1) — the lack of viscosity during the flow of helium through a thin capillary or slit.

All these properties, including the fact that helium exists as a liquid right down to absolute zero, obviously cannot be explained by the classical theory and are connected with quantum phenomena.

L. Tisza (2) suggested that helium II should be considered as a degenerated ideal Bose gas. He suggested that the atoms found in the normal state (a state of zero energy) move through the liquid without friction. This point of view, however, cannot be considered as satisfactory. Apart from the fact that liquid helium has nothing to do with an ideal gas, atoms in the normal state would not behave as «superfluid». On the contrary, nothing could prevent atoms in a normal state from colliding with excited atoms, *i. e.* when moving through the liquid they would experience a friction and there would be no superfluidity at all. In this way the explanation advanced by Tisza not only has no foundation in his suggestions but is in direct contradiction to them.

1. The quantization of the motion of liquids

An arbitrary system of interacting particles (a liquid) can be described in classical theory by means of the density ρ and the flow of the mass $\mathbf{j}$, which are determined in the following manner. Let $\mathbf{R}$ be the radius-vector of an arbitrary point in space and $\mathbf{r}_a$ — the radius-vector of a particle with a mass m_a. Then ρ is determined as

$$\rho=\sum_a m_a \delta(\mathbf{r}_a-\mathbf{R}), \qquad (1.1)$$

δ being the three-dimensional δ-function and the summation is extended over all particles in the system. The volume-integral $\int \rho\, dV$ gives the total mass of the system. Similarly the density $\mathbf{j}$ of the flow of the mass in determined as

$$\mathbf{j} = \sum_\alpha m_\alpha \mathbf{v}_\alpha \delta(\mathbf{r}_\alpha - \mathbf{R}) = \sum_\alpha \mathbf{p}_\alpha \delta(\mathbf{r}_\alpha - \mathbf{R})$$

($\mathbf{v}_\alpha$, $\mathbf{p}_\alpha$ are the velocity and the momentum of the particle m_α).

It must be emphasized that in such a description of a liquid there is no averaging in that sense in which it is done in statistics. This description proceeds from the microscopic picture as all the particles possess (at a given moment) definite coordinates $\mathbf{r}_\alpha$ and velocities $\mathbf{v}_\alpha$.

When passing over to the quantum theory ρ and $\mathbf{j}$ must be regarded as certain operators the form of which must be determined. For the sake of simplicity suppose that the system consists of one particle only. Then the classical density is $\rho = m\delta(\mathbf{r} - \mathbf{R})$. The operator ρ must be determined in such a way that its mathematical expectation $\int \psi(\mathbf{r})^* \rho\psi(\mathbf{r}) dV$ [$\psi(\mathbf{r})$ being the wave function of the particle] equals the density of the mass at the point $\mathbf{R}$, *i. e.* $m|\psi(\mathbf{R})|^2$. From this it follows that the operator ρ must have the same form $\rho = m\delta(\mathbf{r} - \mathbf{R})$ and in the case of an arbitrary system of particles—correspondingly the form (1,1).

The classical density of the flow for one particle is $\mathbf{j} = \mathbf{p}\delta(\mathbf{r} - \mathbf{R})$. It is easy to see that the corresponding operator is

$$\mathbf{j} = \frac{1}{2}[\mathbf{p}\delta(\mathbf{r} - \mathbf{R}) + \delta(\mathbf{r} - \mathbf{R})\mathbf{p}],$$

where $\mathbf{p}$ is the usual operator of momentum:

$$\mathbf{p} = \frac{\hbar}{i}\nabla$$

(∇ denotes the differentiation with respect to $\mathbf{r}$). Actually the mathematical expectation of $\mathbf{j}$ is

$$\int \psi(\mathbf{r})^* \mathbf{j}\psi(\mathbf{r}) dV = \frac{\hbar}{2i}\int \psi^* \nabla\delta(\mathbf{r} - \mathbf{R})\psi\, dV + + \frac{\hbar^*}{2i}\int \psi^* \delta(\mathbf{r} - \mathbf{R})\nabla\psi\, dY$$

or, integrating in the first term in parts:

$$\int \psi^* \mathbf{j}\,\psi\, dV = -\frac{\hbar}{2i}\int \psi\delta(\mathbf{r} - \mathbf{R})\nabla\psi^*\, dV + + \frac{\hbar}{2i}\int \psi^*\delta(\mathbf{r} - \mathbf{R})\nabla\psi\, dV = = \frac{\hbar}{2i}\{\psi^*(\mathbf{R})\nabla\psi(\mathbf{R}) - \psi(\mathbf{R})\nabla\psi^*(\mathbf{R})\},$$

i. e. exactly what it ought to be. For the arbitrary system we have, similarly

$$\mathbf{j} = \frac{1}{2}\sum_\alpha [\mathbf{p}_\alpha\delta(\mathbf{r}_\alpha - \mathbf{R}) + \delta(\mathbf{r}_\alpha - \mathbf{R})\mathbf{p}_\alpha], \quad (1,2)$$

$$\mathbf{p}_\alpha = \frac{\hbar}{i}\nabla_\alpha.$$

We now determine the commutation rules. For the density ρ we obviously have

$$\rho_1\rho_2 - \rho_2\rho_1 = 0 \quad (1,3)$$

[ρ_1, ρ_2 denote $\rho(\mathbf{R}_1)$, $\rho(\mathbf{R}_2)$, respectively].

For the sake of brevity let us consider only one term from each of the sums (1,1) and (1,2) when determining the commutation rules, as the operators corresponding to different particles commute with each other. To determine the commutation of ρ with $\mathbf{j}$ we write

$$\mathbf{j}_1\rho_2 - \rho_2\mathbf{j}_1 = \frac{m\hbar}{2i}\{[\nabla\delta(\mathbf{r} - \mathbf{R}_1) + + \delta(\mathbf{r} - \mathbf{R}_1)\nabla]\delta(\mathbf{r} - \mathbf{R}_2) - - \delta(\mathbf{r} - \mathbf{R}_2)[\nabla\delta(\mathbf{r} - \mathbf{R}_1) + \delta(\mathbf{r} - \mathbf{R}_1)\nabla].$$

To simplify the expression on the right-hand side we note that the operators of the form

$$\delta(\mathbf{r} - \mathbf{R}_1)\nabla\delta(\mathbf{r} - \mathbf{R}_2)$$

can be transformed in the following way:

$$\delta(\mathbf{r} - \mathbf{R}_1)\nabla\delta(\mathbf{r} - \mathbf{R}_2) = \delta(\mathbf{r} - \mathbf{R}_1) \times \times \left(\nabla\delta(\mathbf{r} - \mathbf{R}_2)\right) + \delta(\mathbf{r} - \mathbf{R}_1)\delta(\mathbf{r} - \mathbf{R}_2)\nabla,$$

where in the first term $\left(\nabla\delta(\mathbf{r} - \mathbf{R}_2)\right)$ denotes simply a gradient of the δ-function, *i. e.* ∇ is no longer an operator. Owing to the presence of the factor $\delta(\mathbf{r} - \mathbf{R}_1)$ in this term one can write $\left(\nabla\delta(\mathbf{R}_1 - \mathbf{R}_2)\right)$ instead of $\left(\nabla\delta(\mathbf{r} - \mathbf{R}_2)\right)$ In this way

$$\delta(\mathbf{r} - \mathbf{R}_1)\nabla\delta(\mathbf{r} - \mathbf{R}_2) = \delta(\mathbf{r} - \mathbf{R}_1) \times \times \left(\nabla\delta(\mathbf{R}_1 - \mathbf{H}_2)\right) + \delta(\mathbf{r} - \mathbf{R}_1)\delta(\mathbf{r} - \mathbf{R}_2)\nabla.$$

Similarly

$$\nabla\delta(\mathbf{r}-\mathbf{R}_1)\,\delta(\mathbf{r}-\mathbf{R}_2)=\delta(\mathbf{r}-\mathbf{R}_2)\times$$
$$\times\nabla\delta(\mathbf{r}-\mathbf{R}_1)+\delta(\mathbf{r}-\mathbf{R}_1)\left(\nabla\delta(\mathbf{R}_1-\mathbf{R}_2)\right).$$

The result is

$$\mathbf{j}_1\rho_2-\rho_2\mathbf{j}_1=\frac{\hbar}{i}m\delta(\mathbf{r}-\mathbf{R}_1)\nabla\delta(\mathbf{R}_1-\mathbf{R}_2)$$

or for an arbitrary system

$$\mathbf{j}_1\rho_2-\rho_2\mathbf{j}_1=\frac{\hbar}{i}\rho_1\nabla\delta(\mathbf{R}_1-\mathbf{R}_2). \quad (1,4)$$

(It makes no difference whether we write ρ_1 or ρ_2 on the right-hand side in view of the presence of the δ-function of $\mathbf{R}_1-\mathbf{R}_2$.)

In a similar way the commutation rules between the components of the vector $\mathbf{j}$ with each other can be obtained. The calculation in this case is longer and we will not enter into it here.

We introduce the operator $\mathbf{v}$ of the velocity of the liquid according to

$$\mathbf{j}=\frac{1}{2}(\rho\mathbf{v}+\mathbf{v}\rho), \quad (1,5)$$

$$\mathbf{v}=\frac{1}{2}\left(\frac{1}{\rho}\mathbf{j}+\mathbf{j}\frac{1}{\rho}\right). \quad (1,6)$$

It will be more convenient to use the operator $\mathbf{v}$ instead of the operator of the flow $\mathbf{j}$.

For the commutation rule of ρ with $\mathbf{v}$ we have

$$\mathbf{v}_1\rho_1-\rho_2\mathbf{v}_1=\frac{1}{2}\left(\frac{1}{\rho_1}\mathbf{j}_1+\mathbf{j}_1\frac{1}{\rho_1}\right)\rho_2-$$
$$-\frac{1}{2}\rho_2\left(\frac{1}{\rho_1}\mathbf{v}_1+\mathbf{v}_1\frac{1}{\rho_1}\right)=$$
$$=\frac{1}{2\rho_1}(\mathbf{j}_1\rho_2-\rho_2\mathbf{j}_1)+\frac{1}{2}(\mathbf{j}_1\rho_2-\rho_2\mathbf{j}_1)\frac{1}{\rho_1}$$

or, on inserting (1,4):

$$\mathbf{v}_1\rho_2-\rho_2\mathbf{v}_1=\frac{\hbar}{i}\nabla\delta(\mathbf{R}_1-\mathbf{R}_2). \quad (1,7)$$

The commutation rules for the components of $\mathbf{v}$ are found to be

$$\mathbf{v}_{1i}\mathbf{v}_{2k}-\mathbf{v}_{2k}\mathbf{v}_{1i}=$$
$$=\frac{\hbar}{i}\delta(\mathbf{R}_1-\mathbf{R}_2)\frac{1}{\rho_1}(\operatorname{rot}\mathbf{v})_{ik}, \quad (1,8)$$

where $(\operatorname{rot}\mathbf{v})_{ik}$ denotes the difference

$$\frac{\partial v_k}{\partial x_i}-\frac{\partial v_i}{\partial x_k}.$$

Further on we shall also need the commutation rule between ρ and $\operatorname{rot}\mathbf{v}$. By applying the operation rot (with a differentiation with respect to coordinates $\mathbf{R}_1$) to both sides of the equation (1,7) we get

$$\operatorname{rot}\mathbf{v}_1\cdot\rho_2-\rho_2\cdot\operatorname{rot}\mathbf{v}_1=0. \quad (1,9)$$

It is easy to see that by applying the formulae obtained to the macroscopic movement of the liquid we get, as required, the usual hydrodynamic equations written in an operational form. The energy of a unit volume of a classical liquid considered macroscopically is

$$\frac{\rho v^2}{2}+\rho\varepsilon(\rho),$$

where $\varepsilon(\rho)$ is the internal energy of a unit mass of the liquid. It is supposed that the energy ε depends only on the density ρ of the liquid; this corresponds to the macroscopic character of the consideration and is connected with a statistical averaging. For a microscopic investigation this supposition is, of course, invalid.

The corresponding quantum operator is*

$$\frac{\mathbf{v}\rho\mathbf{v}}{2}+\rho\varepsilon(\rho).$$

The Hamiltonian H of the liquid is an integral over the volume

$$H=\int\left\{\frac{\mathbf{v}\rho\mathbf{v}}{2}+\rho\varepsilon(\rho)\right\}dV. \quad (1,10)$$

For the derivative of the density ρ with respect to time one has

$$\dot\rho=\frac{i}{\hbar}(H\rho-\rho H).$$

We shall denote temporarily the coordinates of the point at which ρ is taken by the index 1, and the coordinates of the variable point in the region of integration in (1,10) by the index 2. Then

$$\dot\rho_1=\frac{i}{\hbar}\int\left\{\frac{1}{2}[\mathbf{v}_2\rho_2\mathbf{v}_2\rho_1-\rho_1\mathbf{v}_2\rho_2\mathbf{v}_2]+\right.$$
$$\left.+[\rho_2\varepsilon(\rho_2)\rho_1-\rho_1\rho_2\varepsilon(\rho_2)]\right\}dV_2.$$

---

* The operator $\frac{\mathbf{v}\rho\mathbf{v}}{2}$ can also be written in the form

$$\frac{\rho\mathbf{v}^2+\mathbf{v}^2\rho}{2}.$$

In view of (1,3) the second term under the sign of the integration vanishes, and the first can be written as

$$\frac{1}{2}[\mathbf{v}_2\rho_2(\mathbf{v}_2\rho_1-\rho_1\mathbf{v}_2)+(\mathbf{v}_2\rho_1-\rho_1\mathbf{v}_2)\rho_2\mathbf{v}_2]$$

or, by introducing (1,7):

$$\cdot\frac{\hbar}{2i}\nabla\delta(\mathbf{R}_2-\mathbf{R}_1)(\mathbf{v}_2\rho_2+\rho_2\mathbf{v}_2)=$$
$$=\frac{\hbar}{i}\nabla\delta(\mathbf{R}_2-\mathbf{R}_1)\,\mathbf{j}_2.$$

In this way

$$\dot{\rho}_1=\int\nabla\delta(\mathbf{R}_2-\mathbf{R}_1)\,\mathbf{j}_2\,dV_2=$$
$$=-\int\delta(\mathbf{R}_2-\mathbf{R}_1)\operatorname{div}\mathbf{j}_2dV_2=-\operatorname{div}\mathbf{j}_1,$$

i. e. we come to the continuity equation in operational form:

$$\frac{\partial\rho}{\partial t}+\operatorname{div}\frac{\rho\mathbf{v}+\mathbf{v}\rho}{2}=0. \qquad (1,11)$$

In a similar way the derivative:

$$\dot{\mathbf{v}}=\frac{i}{\hbar}(H\mathbf{v}-\mathbf{v}H),$$

can be calculated, which brings us to the equation

$$\frac{\partial v_l}{\partial t}+\frac{1}{2}\left(v_k\frac{\partial v_l}{\partial x_k}+\frac{\partial v_l}{\partial x_k}v_k\right)=$$
$$=-\frac{1}{\rho}\frac{\partial}{\partial x_l}\frac{d\varepsilon}{d\rho}, \qquad (1,12)$$

i. e. Euler's equation in an operational form ($d\varepsilon/d\rho$ is the pressure p of the liquid).

It must be again emphasized that the equations (1,11), (1,12) are less general than the commutation rules (1,3)—(1,9), which are also valid for an exact, microscopical investigation of the liquid.

2. The energy spectrum of a quantum liquid

In the classical hydrodynamics of ideal liquids it is shown that if at a certain moment of time, the motion is potential (rot $\mathbf{v}=0$) in the whole volume of the liquid, it will be potential for all other moments of time (Lagrange's theorem). It appears that this classical theorem finds its analogy in quantum hydrodynamics.

According to the commutation rules (1,9), rot $\mathbf{v}$ always commutes with the density ρ. The components of rot $\mathbf{v}$, however, do not commute, generally speaking, either with each other or with the components of velocity $\mathbf{v}$ [when the operation rot is applied to the equation (1, 8) the right-hand side does not vanish]. Therefore, rot $\mathbf{v}$ does not, generally speaking, commute with the Hamiltonian, *i. e.* is not conserved.

An exception is the case when over the whole volume of the liquid rot $\mathbf{v}=0$. In this case we have zero in the right-hand side of (1,8) and rot $\mathbf{v}$ commutes with ρ and $\mathbf{v}$ and, therefore, also with the Hamiltonian*.

In this way rot $\mathbf{v}$ is conserved if it is zero. In other words, a quantum liquid always possesses stationary states in which rot $\mathbf{v}$ equals zero over the whole volume of the liquid. Such a state might be called, by analogy to classical hydrodynamics, a state of potential motion of the liquid.

Concerning these results an analogy can be made with the angular momentum $\mathbf{M}$ in quantum mechanics. The commutation of two components of $\mathbf{M}$ with each other leads to the third component of $\mathbf{M}$, with the result that all the components of $\mathbf{M}$ commute with each other if they are all equal to zero. It is also known that there exist no states with an infinitely small angular momentum, its first non-zero eigenvalues are of the order of $\hbar$. This is a consequence of the fact that the commutation rules are inhomogeneous — their left-hand sides are quadratic in $\mathbf{M}$ and the right-hand sides are linear.

A similar statement can be advanced concerning rot $\mathbf{v}$ in quantum hydrodynamics. Namely, no states can exist in which rot $\mathbf{v}$ would be non-zero, but arbitrarily small over the whole volume of the liquid. In other words, between the

* Not only with the Hamiltonian (1, 10), but also with all other functions containing ρ, $\mathbf{v}$ and their derivatives of any order with respect to the coordinates.

states of the potential (rot $\mathbf{v} = 0$) and vortex (rot $\mathbf{v} \neq 0$) motions of a quantum liquid there is no continuous transition.

From this the principle features of the energy spectrum of a liquid directly follow. The presence of a gap between the states of the potential and vortex motions means that between the lowest energy levels of vortex and potential motions a certain finite energy interval must exist. As to the question which of these two levels lies lower, apparently both cases are logically possible. It will be shown below that we get the phenomenon of superfluidity if we suppose that the normal level of the potential motions lies lower than the normal level of vortex motions. Hence we must suppose that this very case exists in liquid helium. It must be remarked, however, that, as only one quantum liquid exists, liquid helium, the question as to whether such a distribution of the levels and hence the property of superfluidity is a general property of a quantum liquid cannot be solved experimentally.

This brings us to the following picture of the distribution of the levels in the energy spectrum of liquid helium (it must be emphasized that we do not here refer to the levels for single helium atoms but to the levels corresponding to the states of the whole liquid). This spectrum is made up of two superimposed continuous spectra. One of them corresponds to the potential motions and the other — to vortex motions. The lowest level of the vortex spectrum is situated above the lowest level of the potential spectrum, this latter level being the normal unexcited state of a liquid; the energy interval between these two levels we denote by Δ.

The value of the energy gap Δ cannot be calculated exactly. Its order of magnitude is

$$\Delta \sim \frac{m^5}{\rho^{2/3}\hbar^2} \qquad (2,1)$$

(m being the mass of the helium atom and ρ — the density of the liquid). This is the only quantity of the dimension of energy which can be built up from m, ρ and $\hbar$. This gives numerically $\Delta/k \sim 1°$, *i. e.* Δ, as was expected, is of the order of kT_λ, T_λ being the temperature of the λ-point of helium [cf. (3,8)].

Consider an excited level which is situated not too high above the beginning of the spectrum (vortex or potential one).

Every weakly excited state can be considered as an aggregate of a number of single «elementary excitations». As far as the excited levels of the potential spectrum are concerned, the potential internal motions of the liquid are longitudinal waves, *i. e.* these motions are sound waves. Therefore, the corresponding elementary excitations are simply sound quanta, *i. e.* phonons. The energy of the phonons is known to be a linear function of their momentum p:

$$\varepsilon = cp, \qquad (2,2)$$

c being the velocity of sound. Thus, at the beginning of the potential spectrum the energy is proportional to the first power of the momentum.

An «elementary excitation» of the vortex spectrum might be called a «roton»* Those special reasons which stipulate a linear dependence of ε on p for phonons do not exist for rotons. For small momenta $\mathbf{p}$ the energy of the roton can be simply expanded in powers of $\mathbf{p}$; in view of the isotropy of the liquid the expansion of the scalar ε in powers of the vector $\mathbf{p}$ only contains members of the even powers, so one may write:

$$\varepsilon = \Delta + \frac{p^2}{2\mu}, \qquad (2,3)$$

where μ is an «effective mass» of the roton [in (2,2) and (2,3) the energy is measured from the normal state]**.

If the number of phonons and rotons (per unit volume of the liquid) is not large, their aggregate can be regarded as a mixture of two ideal gases — a phonon gas and a roton gas. It is known that the phonon gas obeys Bose statistics. As to the rotons,

* This name was suggested by I. E. Tamm.

** In a recent paper A. Bijl (3) investigated the properties of the energy spectrum of a liquid and came to the conclusion that there must be an energy gap between the normal and all excited states. This result does not seem to be plausible as it would mean, in particular, the impossibility of the propagation of sound waves with small frequencies in liquids.

they too probably obey Bose statistics. It must, however, be remarked that inasmuch as the energy of a roton always contains a quantity Δ large compared with kT (at low temperatures only when the aggregate of rotons can be treated as a gas) the difference between the Bose and Fermi statistics is not essential and one can use Boltzmann's distribution for the rotons.

3. The heat capacity of helium II

From the properties of the energy spectrum one can come to certain conclusions concerning the temperature dependence of the heat capacity of helium II.

At sufficiently low temperatures, when the abnormal behaviour of the heat capacity near the λ-point is no longer essential, the heat capacity may be considered as consisting of two parts — a phonon and a roton part. The first (C_{ph}) is no other than the Debye heat capacity proportional to the cube of the temperature. This part of the heat capacity can be calculated exactly* as the Debye temperature can be calculated from the data obtained ([4]) for the compressibility of helium from which, in its turn, the velocity of sound in helium can be calculated (it turns out to be 250 m/sec.). According to the known formula we get, for the heat capacity per 1 g of helium

$$C_{ph}=4.4\cdot 10^{-3}T^3\,\frac{\text{cal}}{\text{deg}\cdot\text{g}}\,. \tag{3,1}$$

The roton part of the heat capacity (C_r) has, owing to the presence of the energy gap Δ between the normal and vortex levels, an exponential temperature dependence. The coefficient before the exponential factor can be expressed through Δ and the effective mass μ of the roton. For the temperatures considered one can, as has already been pointed out, apply Boltzmann's distribution to the roton gas. The free energy of the gas with an ungiven number of particles (the number of particles in the roton gas is itself a temperature function determined from the condition of the minimum of the free energy) is

$$F_r=-kTV\int e^{-\varepsilon/kT}\frac{d\tau_p}{(2\pi\hbar)^3}$$

$$(d\tau_p=dp_x dp_y dp_z)$$

(V is the volume). Below we shall refer the free energy to 1 g of helium corresponding to which we shall put $V=1/\rho$. (ρ being the helium density), so that

$$F_r=-\frac{kT}{\rho(2\pi\hbar)^3}\int e^{-\varepsilon/kT}d\tau_p. \tag{3,2}$$

The number of rotons is

$$N_r=\frac{1}{\rho(2\pi\hbar)^3}\int e^{-\varepsilon/kT}d\tau_p. \tag{3,3}$$

Putting $\varepsilon=\Delta+p^2/2\mu$ we get:

$$F_r=-\frac{kT}{\rho}\left(\frac{\mu kT}{2\pi\hbar^2}\right)^{3/2}e^{-\Delta/kT}; \tag{3,4}$$

$$N_r=\frac{1}{\rho}\left(\frac{\mu kT}{2\pi\hbar^2}\right)^{3/2}e^{-\Delta/kT}. \tag{3,5}$$

Thus the entropy is

$$S_r=-\frac{\partial F_r}{\partial T}=$$

$$=\frac{1}{\rho}\left(\frac{\mu k}{2\pi\hbar^2}\right)^{3/2}T^{1/2}\Delta e^{-\Delta/kT}\left(1+\frac{5kT}{2\Delta}\right), \tag{3,6}$$

and the heat capacity is

$$C_r=T\frac{\partial S_r}{\partial T}=\frac{1}{\rho}\left(\frac{\mu}{2\pi\hbar^2}\right)^{3/2}\frac{k^{1/2}\Delta^2}{T^{1/2}}e^{-\Delta/kT}\times$$

$$\times\left[1+3\frac{kT}{\Delta}+\frac{15}{4}\left(\frac{kT}{\Delta}\right)^2\right]. \tag{3,7}$$

If we use the data obtained by Keesom and Miss Keesom ([5]) for the heat capacity in the interval 1.2 — 1.5° K from the formulae derived we get for μ and Δ the values:

$$\frac{\Delta}{k}=8-9^\circ, \qquad \mu=7-8\,m_{He} \tag{3,8}$$

(m_{He} is the mass of a helium atom).

At sufficiently low temperatures C_r must become smaller than C_{ph}. It must, however, be remarked that the calculated phonon heat capacity according to (3,1) is found to be at 1.2° K about 22 times less than its experimental value. In this way at temperatures even twice as low as the λ-point the roton gas still plays a predominant rôle in the heat capacity.

* This calculation had already been made in 1940 by A. Migdal whom I wish to thank for informing me of his results.

Up to the present time the heat capacity at lower temperatures has only been measured by B. Bleaney and F. Simon ([6]) (in a temperature interval from 0.25 to 0.8° K). Their results still give a value ca. 10 times larger than that given by the formula (3,1). On the other hand, Bleaney and Simon found it possible to describe the heat capacity in the temperature interval 0.25 to 0.8°K by a T^3- law. From this it would follow that the surplus part of the heat capacity over C_{ph}, *i. e.* the heat capacity C_r, also obeys the T^3- law, whereas according to the theory one would expect it to have an exponential temperature dependence. It must be mentioned, however, that Bleaney and Simon did not publish details of their measurements and so the possibility is not excluded that their results ought to be corrected. On the other hand, P. L. Kapitza's ([9]) measurements show that the value of the entropy at 1.33°K calculated from the formula (3,6) with μ and Δ from (3,8) agree well with experiment, whereas the entropy value calculated from the data of Bleaney and Simon does not agree with Kapitza's results.

4. Superfluidity of helium II at absolute zero

It will now be shown that the superfluidity of helium II is a consequence of the properties of the energy spectrum described above. Consider first liquid helium at absolute zero. At this temperature the liquid is in its normal, unexcited state. Let us consider the helium when flowing through a capillary at a constant velocity **V**. The presence of viscosity would be manifested by the fact that owing to the friction between the liquid and the walls of the capillary and the friction inside the liquid itself the kinetic energy of the movement of the liquid would be dissipated and the flow gradually slowed down.

It is convenient to consider this flow in the coordinate system moving together with the helium. In this system the helium is at rest and the walls of the capillary move with a velocity $-\mathbf{V}$. In the presence of viscosity the helium at rest should also begin to move. Physically it is obvious that the interaction of the liquid with the moving walls of the tube cannot cause a motion of the whole liquid at once. The motion must begin with the excitation of the inner movements in the layers of liquid close to the walls of the tube, *i. e.* with the excitation of rotons and phonons in the liquid.

Let us suppose that a phonon can be excited in the liquid. Then the energy E_0 of the liquid (in the coordinate system in which it was initially at rest) is equal to the energy ε of the phonon (we will agree to take the energy of the normal state as zero) and the momentum $\mathbf{P}_0$ of the liquid is the momentum $\mathbf{p}$ of the phonon. As $\varepsilon = cp$ we have:

$$E_0 = cp, \quad \mathbf{P}_0 = \mathbf{p}.$$

We will now return to the coordinate system in which the capillary is at rest. According to the known formulae for the transformation of energy and momentum in classical mechanics we have, for the energy E and the momentum $\mathbf{P}$ in this system

$$E = E_0 + \mathbf{P}_0\mathbf{V} + \frac{MV^2}{2}, \qquad \mathbf{P} = \mathbf{P}_0 + M\mathbf{V},$$

M being the mass of the liquid. Inserting E_0 and $\mathbf{P}_0$ we find for the energy

$$E = cp + \mathbf{pV} + \frac{MV^2}{2}. \qquad (4,1)$$

The term $MV^2/2$ is the initial kinetic energy of the flowing helium, the expression $cp + \mathbf{pV}$ being the change of energy owing to the excitation of the phonon. This change must be negative as the energy of the flowing liquid must decrease:

$$cp + \mathbf{pV} < 0.$$

To fulfil this condition the absolute value of the velocity **V** must be, in any case, greater than the velocity of sound

$$V > c. \qquad (4,2)$$

At smaller velocities the interaction with the walls of the tube cannot give rise to an excitation of phonons.

An analogous consideration can be applied to the excitation of a roton. In this case

$$E_0 = \varepsilon = \Delta + \frac{p^2}{2\mu},$$

and instead of (4,1) we get:

$$E = \Delta + \frac{p^2}{2\mu} + \mathbf{V}_p + \frac{MV^2}{2}. \qquad (4,3)$$

For $E < MV^2/2$ it is at any rate necessary that

$$\Delta + \frac{p^2}{2\mu} - p < 0,$$

i. e.

$$V > \frac{p}{2\mu} + \frac{\Delta}{p}.$$

The expression on the right-hand side has a minimum at $p^2 = 2\mu\Delta$ so that for the excitation of rotons the velocity must be greater than

$$V > \sqrt{\frac{2\Delta}{\mu}}. \qquad (4,4)$$

In this way we see that neither phonons nor rotons can be excited if the velocity of flow in helium II is not too large. This means that the flow of the liquid does not slow down, *i. e.* helium II discloses the phenomenon of superfluidity*.

It must be remarked that already the reasons given above are enough to make the superfluidity vanish at sufficiently large velocities. We leave aside the question as to whether superfluidity disappears at smaller velocities for some other reason [the velocity limit obtained from (4,2) is very large — the velocity of sound in helium equals 250 m/sec.; (4,4) gives a value only several times lower].

5. Helium II at temperatures above absolute zero

We now consider helium at non-zero temperatures. Helium is not then in its normal state and at sufficiently low temperatures its excitation can be considered as a gas consisting of phonons and rotons. The results of § 4 still remain valid as the circumstance that liquid helium is initially in the normal state was not used directly in our derivation. Therefore, new phonons and rotons cannot be excited owing to the movement of the helium relative to the walls of the vessel also at temperatures greater than 0° K. It is necessary, however, to find out how the phonons and rotons that are already in the liquid will manifest themselves.

To investigate this problem we shall consider helium in an axial-symmetric vessel rotating round its axis at a constant angular velocity Ω. We go over to the coordinate system rotating together with the vessel. In this system the vessel is at rest, *i. e.* the external conditions in which the liquid is found are stationary. Therefore, the Gibbs distribution here holds, *i. e.* the probability of finding helium in an excited state is determined by the formula

$$\text{const} \cdot e^{-E'/kT},$$

E' being the energy of the excited state of the liquid in the rotating coordinate system. It is known that the energy E' of a body in the rotating coordinate system is connected with the energy in the stationary system by means of

$$E' = E - \mathbf{M}\Omega,$$

where $\mathbf{M}$ is the angular momentum of the body in a given excited state (in the normal state $\mathbf{M} = 0$). Thus, the Gibbs distribution in the stationary coordinate system is

$$\text{const} \cdot e^{-(E + \mathbf{M}\Omega)/kT}. \qquad (5,1)$$

For the sake of clarity let us suppose that the temperature is low enough so that one can speak of an ideal gas of phonons and rotons in helium. Then the energy E and angular momentum $\mathbf{M}$ of the excited state can be written as

$$E = \sum \varepsilon, \quad \mathbf{M} = \sum \mathbf{m}, \qquad (5,2)$$

ε and $\mathbf{m}$ being the energy and angular momentum respectively of single phonons and rotons.

As is well known, if we substitute $E = \sum \varepsilon$ in the Gibbs distribution $e^{-E/kT}$ we can pass over to the distribution for

* It must be mentioned that for an ideal gas $\Delta = 0$ and, therefore, even at absolute zero it would not disclose the phenomenon of superfluidity at any velocities of the flow contrary to Tisza's suggestion.

single «particles» of the gas, in the given case to the Bose distribution:

$$\frac{1}{e^{\varepsilon/kT}-1},$$

for phonons and probably for rotons (which particular distribution is not, however, essential to what will come).

Similarly, by inserting (5,2) into (5,1) we get in the same way the same distribution for phonons and rotons in the rotating vessel with the only difference that instead of ε enters $\varepsilon - \mathbf{m}\Omega$, so that the Bose distribution will take the form:

$$\frac{1}{e^{(\varepsilon-\Omega\mathbf{m})/kT}-1}.$$

But such a distribution is no other than the distribution for the gas rotating as a whole (at an angular velocity Ω). And so we come to the result that during the rotation of the vessel containing helium II a statistical equilibrium is established which differs from the equilibrium in a vessel at rest only in that the photon and roton gas rotates together with the vessel as if it was carried along by the walls of the vessel.

If, from the above distribution one calculates the angular momentum of the helium in the rotating vessel at a given temperature, *i. e.* the quantity:

$$\overline{\mathbf{M}} = \frac{1}{(2\pi\hbar)^3}\int\int\frac{\mathbf{m}d\tau_p\,dV}{e^{(\varepsilon-\Omega\mathbf{m})/kT}-1}, \quad (5,3)$$

then at absolute zero, *i. e.* when phonons and rotons are completely absent we would obviously get zero. At higher temperatures the angular momentum will be non-zero but the momentum of inertia (*i. e.* the proportionality factor between $\mathbf{M}$ and Ω) will be at sufficiently low temperatures much lower than the usual one (which corresponds to the rotation of the total mass of the liquid with the vessel).

This brings ut to the fundamental result that when the walls of the vessel are in motion only a part of the liquid helium mass is carried along by them, the other part «remaining stationary». And so we might regard liquid helium as if it was a mixture of two liquids, one of which is superfluid having no viscosity and not carried along by the walls of the vessel and the other — a «normal» one which is carried along by the moving walls and behaves itself like a normal liquid. It is most essential that «there is no friction» between these two liquids moving through each other, *i. e.* that there is no transfer of momentum from one to the other. Actually we get this relative motion when considering the statistical equilibrium in an uniformly rotating vessel. But if there can be some sort of relative motion in the state of statistical equilibrium it means that it cannot be accompanied by friction.

It must be stressed that when we look upon helium as a mixture of two liquids it is no more than a method of expression, convenient for describing phenomena which take place in helium II. Like every description of quantum phenomena in classical terms it is not quite adequate. Actually one must say that in a quantum luquid two movements can exist simultaneously each of which is connected with its own «effective mass» (so that the sum of both these masses equals the total real mass of the liquid). One of these movements is normal, *i. e.* possesses the same properties as the movements of usual liquids; the other is superfluid. Both these motions take place without a transfer of the momentum from one to the other. We particularly emphasize that there is no division of the real particles of the liquid into «superfluid» and «normal» ones here. In a certain sense one can speak of «superfluid» and «normal» masses of liquid as of masses connected with the two simultaneously possible movements, but this by no means signifies the possibility of a real division of the liquid into two parts.

Keeping in mind these reservations as to the real nature of the phenomena taking place in helium II it is still convenient to use the terms «superfluid» and «normal» liquids as a short and convenient method of describing these phenomena and in the further discussion we shall do so.

During the rotation of the vessel containing helium the superfluid part remains stationary as has already been pointed out. It can be said that the superfluid liquid is not capable of rotation. Mathematically this means that the rotor of the velocity of the superfluid motion equals zero. Hence the motion of the superfluid liquid is always potential*. As to the normal part of the liquid it can accomplish a potential or a vortex motion.

This brings us in particular to the following interesting property of the motion of helium II. From hydrodynamics we know that during the potential flow of a liquid no pressure is exerted on a body immersed in the liquid (the so called Euler's paradox). Therefore, the superfluid part will not, during its motion, exert any pressure on a body immersed in helium II; the body will only be influenced by the normal part of the liquid.

At low temperatures, when the excitation is not too great the normal part of the liquid is an aggregate of phonons and rotons. Therefore, the viscosity of the normal liquid can be calculated in the same way as that of a mixture of phonon and roton gases. Even at not very low temperatures, *i. e.* at such temperatures when the number of rotons is large compared to the number of phonons, we cannot completely disregard the phonons and calculate the viscosity as the viscosity of the roton gas alone because the «length of the free path» of phonons is, as can be shown, large as compared to that of the rotons.

The most important parameter which determines the properties of helium at every given temperature is the ratio of the masses of the superfluid and normal parts of the liquid. We introduce the density ρ_n of the normal liquid and the density ρ_s of the superfluid; the sum $\rho = \rho_n + \rho_s$ is the total true density of the liquid.

At absolute zero the ratio ρ_n/ρ equals zero. As the temperature rises it increases until it becomes equal to unity after which, of course, it will remain constant. The temperature at which ρ_n/ρ becomes unity is the point of transition from helium II into helium I. In this way the phase transition in helium is connected with the disappearance of the superfluid part of the liquid. This disappearance takes place continuously, *i. e.* ρ_n/ρ becomes unity continuously, without a jump. Therefore, the transition is a phase transition of the second order, *i. e.* a λ-point (it is not accompanied by the absorption or giving out of latent heat). The presence of a jump in the heat capacity is a direct thermodynamical consequence of a phase transition of the second order [cf., *e. g.* (7)].

We will here point out the following possible method for a direct experimental determination of the curve of ρ_n/ρ as a function of the temperature. If the moment of inertia J of the vessel filled with helium II and rotating about its axis is measured the ratio of J to the moment of inertia J_0, calculated on the supposition that the total mass of helium rotates with the vessel, determines the ratio ρ_n/ρ at a given temperature.

At sufficiently low temperatures the ratio ρ_n/ρ can be calculated theoretically. At these temperatures ρ is built up of two independent parts — the effective mass of an ideal gas of phonons and the mass of a roton gas. We will first calculate the phonon part of ρ_n. To do this we shall imagine that the liquid, and with it the phonon gas, moves as a whole with a constant velocity $\mathbf{V}$. As is known, the distribution function for a gas, moving as a whole, is obtained from the distribution function of the stationary gas simply by inserting the quantity $\varepsilon - \mathbf{pV}$ in place of the energy ε of the particle (where $\mathbf{p}$ is the momentum of the particle). Hence the distribution function for the phonons obeying Bose statistics is

$$\frac{1}{(2\pi\hbar)^3}\,\frac{1}{e^{(\varepsilon-\mathbf{pV})/kT}-1}.$$

We now calculate the total momentum $\mathbf{P}$ of the phonon gas (per unit volume). We have:

$$\mathbf{P}=\frac{1}{(2\pi\hbar)^3}\int\frac{\mathbf{p}\,d\tau_p}{e^{(\varepsilon-\mathbf{Vp})/kT}-1}.$$

Suppose that the velocity $\mathbf{V}$ is small and expand the integrated expression in powers of $\mathbf{pV}$. The zero order term vanishes after

* A more strict proof of this statement will be given in § 9 where the analogous questions in superconductors are being investigated.

averaging over the directions of $\mathbf{p}$ and we obtain:

$$\mathbf{P}=-\int \mathbf{p}(\mathbf{p}\mathbf{V})\frac{\partial n_0}{\partial \varepsilon}d\tau_p$$

$$n_0=\frac{1}{(2\pi\hbar)^3}\frac{1}{e^{\varepsilon/kT}-1}.$$

As for the phonons we have $\varepsilon=cp$ and, therefore,

$$c\frac{\mathbf{p}}{p}\frac{\partial n_0}{\partial \varepsilon}=\frac{\partial n}{\partial \mathbf{p}},$$

so we can write

$$\mathbf{P}=-\frac{1}{c}\int \mathbf{p}(\mathbf{p}\mathbf{V})\frac{\partial n_0}{\partial p}d\tau_p.$$

Integrating by parts and averaging over the directions of the momentum $\mathbf{p}$ we get:

$$\mathbf{P}=\frac{4}{3c}\mathbf{V}\int pn_0\,d\tau_p=\frac{4}{3c^2}\mathbf{V}\int \varepsilon n_0\,d\tau_p.$$

But the integral entering this formula is the energy ρE_{ph} per unit volume of the phonon gas, so that

$$\mathbf{P}=\frac{4}{3}\rho\mathbf{V}\frac{E_{\text{ph}}}{c^2}$$

The coefficient at $\mathbf{V}$ is the effective mass of a unit volume of the phonon gas, *i. e.* the phonon part $\rho_n^{(\text{ph})}$ of the density ρ_n:

$$\rho_n^{(\text{ph})}=\frac{4}{3}\frac{E_{\text{ph}}}{c^2}\rho. \qquad (5,4)$$

The energy E_{ph} of the phonon gas, and hence $\rho_n^{(\text{ph})}$ is proportional to T^4.

A similar calculation gives the following expression for the roton part of the mass of the normal liquid at low temperatures (the Boltzmann distribution with $\varepsilon=\Delta+ +p^2/2\mu$ is used for the rotons):

$$\rho_n^{(r)}=N_r\mu\rho \qquad (5,5)$$

[N_r is taken out of (3,4)]. The number N_r of rotons, and hence also $\rho_n^{(r)}$ depend on the temperature exponentially ($\sim e^{-\Delta/kT}$, cf. § 3).

ρ_n/ρ is plotted against the temperature in Fig. 1. Even at 1°K the phonon part of ρ_n is thirty times less than the roton part. The part of the curve in Fig. 1 which is near to the λ-point cannot, of course, be calculated theoretically and is obtained by means of an interpolation.

One may, however, expect that owing to a very rapid increase of ρ_n according to formula (5,5), the value of the temperature of the λ-point itself can be approximately obtained from $\rho_n/\rho=1$ using formula (5,5)

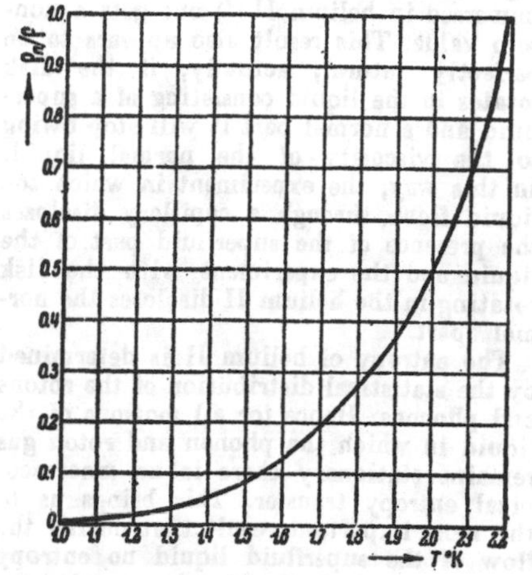

Fig. 1

for ρ_n. Such a calculation gives 2.3°K for the temperature of the λ-point which is in a good agreement with the known value 2.19°K.

6. The flow through capillaries and the heat conductivity of helium II

The ideas developed above allow one to give a satisfactory explanation of a number of results experimentally observed and also to predict a number of new properties which one might expect to find in helium II.

P. L. Kapitza (1) showed that helium II discloses no viscosity when flowing through a capillary or a narrow slit. From the point of view of the theory advanced, this is explained by the fact that when helium II flows out of a vessel through a narrow slit the superfluid part of the liquid flows through the slit disclosing no friction; the normal part remains in the vessel flowing through the slit incomparably slower with a velocity corresponding to its viscosity and the thickness of the capillary.

It is known that when the viscosity is measured by means of observing the damping of the torsional vibrations of a disk immersed in helium II ([8]) one gets a nonzero value. This result also appears to be perfectly natural; actually, if the disk rotates in the liquid consisting of a superfluid and a normal part it will stop owing to the viscosity of the normal liquid. In this way, the experiment in which the liquid flows through a capillary discloses the presence of the superfluid part of the liquid and the experiment with the disk rotating in the helium II discloses the normal part.

The entropy of helium II is determined by the statistical distribution of the rotons and phonons. Hence for all motions of the liquid in which the phonon and roton gas remains stationary there is no macroscopical entropy transfer. This brings us to the most important result that during the flow of the superfluid liquid no entropy transfer takes place. In other words, the motion of the superfluid liquid does not carry heat. From this it follows in its turn that the motion of helium II in which only the superfluid part of the liquid takes part is thermodynamically reversible.

It was shown above that during the flow of helium II through a thin capillary or through a slit we have to deal with just such a flow of the superfluid liquid. Therefore, this flow must be reversible (to be more exact, the thinner the capillary and the less the normal liquid penetrates it the closer we are to a complete reversibility). This was disclosed during the recent ingenious experiments made by P. L. Kapitza ([9]).

As the helium flowing through a thin capillary does not carry heat, we may come to an important conclusion. Namely, liquid helium flowing out of a vessel through a thin capillary must be at a temperature of absolute zero (it would be more true to say at a temperature lower than the temperature of the helium in the vessel and equal to zero only in the ideal case of an infinitely thin capillary).

The known property of helium of forming a moving film on a hard surface is an effect akin to the properties of its flow through a thin capillary. The fact itself that a film is formed is not a peculiar property only to be found in helium II. Films are formed by all liquids wetting a given hard surface. But in ordinary liquids the formation of the film takes place extremely slow owing to the presence of viscosity. The formation and movement of the film of helium II takes place quickly owing to the superfluidity.

The so called thermomechanical effect in helium II, as is known, consists in the fact that when helium flows out of a vessel through a thin capillary a heating is observed in the vessel; on the other hand, at the place where the helium flows into a vessel from a capillary a cooling takes place. The presence of the thermomechanical effect is not in itself peculiar only to helium*, anomalous is only the large value of the effect. The thermomechanical effect in ordinary liquids is an irreversible phenomenon of the same type as the thermoelectric Peltier effect.

An effect of this type must also exist in helium II, however, in this case it is overlapped by a much stronger effect which is specific to helium II having nothing in common with the irreversible phenomena of the type of the Peltier effect. Namely, the heating when the helium flows out through a capillary is simply due to the fact, that this helium does not carry heat, therefore, the heat in the vessel is distributed over a smaller amount of helium. When the helium flows in we find the reverse phenomenon.

It is easy to write what heat Q is absorbed when 1 g of helium flows out of the vessel through a capillary. As the liquid flowing out does not contain phonons and rotons its entropy equals zero. If the helium in the vessel is to remain at its initial temperature T the amount of heat TS (S — entropy of 1 g of helium at a temperature T) must be conveyed to it to compensate for the decrease which takes place per unit mass of the entropy owing to the introduction

* The thermomechanical effect was observed recently in water by B. V. Deriaguin.

of 1 g of helium with zero entropy. This means that if 1 g of helium flows into a vessel containing helium at a temperature ·T, the amount of heat absorbed is

$$Q = TS. \qquad (6,1)$$

In the opposite case when 1 g of helium flows out of the vessel containing helium at a temperature T this amount of heat is given off. The heat Q was recently measured by P. L. K a p i t z a (⁹). The results obtained are found to be in excellent agreement with formula (6,1).

We shall now consider two vessels containing helium II at temperatures T_1 and T_2 and connected to each other by a thin capillary. As the superfluid liquid is able to flow freely along the capillary a mechanical equilibrium of the helium in the two vessels is quickly established. But as, however, the superfluid liquid does not carry heat, the heat equilibrium at which the temperature of the helium in the two vessels equalizes is only established much slower.

The conditions of the mechanical equilibrium of helium can be easily written down by using the·fact that the establishment of this equilibrium takes place, according to what has gone before, at constant entropies S_1 and S_2 of helium in both vessels. If E_1 and E_2 are the internal energies per unit mass of helium at temperatures T_1 and T_2 the conditions of the mechanical equilibrium (i. e. the minimum energy) due to the transfer of the superfluid liquid is

$$\left(\frac{\partial E_1}{\partial N_1}\right)_{S_1} = \left(\frac{\partial E_2}{\partial N_2}\right)_{S_2},$$

N being the number of atoms in 1 g of helium. But the derivative $(\partial E/\partial N)_S$ is the chemical potential ζ. Therefore, the condition of equilibrium is obtained in the form:

$$\zeta(p_1, T_1) = \zeta(p_2, T_2)$$

(p_1, T_1 and p_2, T_2 being the pressure and temperature in the first and second vessel) or

$$\Phi(p_1, T_1) = \Phi(p_2, T_2), \qquad (6,2)$$

where Φ is the thermodynamic potential of 1 g of helium II.

6*

If the pressure p_1 and p_2 are small, we can expand Φ in their powers and remembering that $\partial\Phi/\partial p$ is the specific volume V we get:

$$V\Delta p = \Phi(T_1) - \Phi(T_2) = \int_{T_1}^{T_2} SdT \qquad (6,3)$$

($\Delta p = p_2 - p_1$). If the temperature difference $\Delta T = T_2 - T_1$ is also small we can also expand in powers of ΔT and remarking that $\partial\Phi/\partial T = -S$ we get:

$$\frac{\Delta p}{\Delta T} = \frac{S}{V}. \qquad (6,4)$$

As $S > 0$, and $V > 0$ also $dp/dT > 0$ in agreement with the experiment.

The formulae (6,1) and (6,4) were deduced already by H. L o n d o n (¹⁰) starting from Tisza's ideas, the verbal formulation of which coincides at this point with the theory here advanced. These formulae are fully confirmed by P. L. K a p i t z a 's (⁹) experiments.

We can note that the phenomena described can be regarded in the framework of the theory advanced here as the osmotic phenomena in the «solutions» of phonons and rotons in liquid helium, the narrow capillary or slit playing the rôle of a semipermeable membrane.

Finally, we come to the heat conductivity in helium II. The process of heat transfer in helium II must be represented ·in the following form. Consider the heat conductivity of helium II in a capillary the large value of which was discovered by W. K e e - s o m and Miss K e e s o m (¹¹) and studied in detail by P. L. K a p i t z a (¹²). If the helium is heated at one end of the capillary (or in a bulb soldered to it) two oppositely directed currents arise in helium. The normal liquid carrying the heat flows from the heated to the cold end; the heat which is transferred in this way is quite sufficient to explain the large value of the transfer experimentally observed. The current of the superfluid liquid flows in the opposite direction. The masses of liquid transferred by each current exactly compensate each other so that no real macroscopical flow in the helium takes place.

P. L. Kapitza [12] observed the deflection of a leaf hung in front of the open end of the capillary when the helium in the bulb at the other end of the capillary was heated. P. L. Kapitza suggested that the explanation of this phenomena might be that it was due to the axial flow of the helium from the heated to the cold end occupying practically the whole of the cross-section of the tube; the amount of helium in the tube was supposed to be sustained by means of the reverse flow of helium moving into the capillary along its surface.

According to the quantum representation developed above this phenomenon has a different appearance. The current of superfluid liquid flowing into the capillary exerts no pressure on the leaf (owing to the potentiality of the movement, cf. § 5). On the other hand, the current of normal liquid exerts pressure on the leaf deflecting it in the direction observed. Both these currents go along the whole cross-section of the capillary and there is no more need to introduce the concept of a current of liquid flowing along the walls with anomalous properties.

It must again be emphasized that the concept of a superfluid and normal liquid is only a convenient way of describing the phenomena, but one must in reality speak of two movements which take place simultaneously in one and the same liquid; one of these movements carries heat and the other does not.

We thus come to a peculiar picture; on a body immersed in helium II a force can act in the absence of any movement of the liquid as a whole.

Consider also heat conductivtiy in large volumes of helium. If a body with a heated surface is immersed in helium II again two currents arise in the volume of liquid—the current of normal liquid flowing from the heated surface and an opposite current of the superfluid liquid. The transition of the currents one into the other, *i. e.* the transition of the superfluid liquid into the normal one takes place in the thin layer near the surface. The thickness of this layer must be of the order of the length of the free path of the phonons and rotons forming the normal part of the liquid. Almost the whole temperature fall will obviously take place in this layer and it follows that in the main volume of the liquid we can expect an almost total lack of temperature gradient. Such a distribution of temperature near the heated surface appears to be supported by P. L. Kapitza's [12] experiments.

7. Equations of the macroscopic hydrodynamics of helium II

Starting from the above considerations on the microscopic mechanism of the phenomenon of superfluidity a complete system of hydrodynamic equations can be built which would describe helium II on a macroscopic (phenomenological) way.

The starting point is the fundamental circumstance that the motion of helium II must be described not by one velocity as in ordinary hydrodynamics but by two. One of these is the «superfluid» velocity (denoted by $\mathbf{v}_s$) satisfying the condition

$$\mathrm{rot}\,\mathbf{v}_s = 0. \tag{7,1}$$

On the boundary of a hard surface only the normal component of $\mathbf{v}_s$ becomes zero and not its tangential one corresponding to the fact that the superfluid liquid is not held back by friction against the walls of the vessel. For the «normal» velocity $\mathbf{v}_n$ of the liquid on the boundary with a hard surface the condition $\mathbf{v}_n = 0$ (as in ordinary viscous liquids) must be fulfilled which expresses the fact that the normal liquid is brought to a standstill owing to the friction against the walls.

It turns out that the hydrodynamic equations with the two velocities $\mathbf{v}_s$ and $\mathbf{v}_n$ can be obtained absolutely synonymously starting from the one condition only that they should satisfy all the conservation laws. These equations for the general case of arbitrary velocities are somewhat complicated and we shall not give them here and confine ourselves to a simplified deduction of the equations applicable to the motion with not too large velocities $\mathbf{v}_s$ and $\mathbf{v}_n$.

Let $\mathbf{j}$ be the macroscopic current of the mass of liquid; it is the function of the both velocities $\mathbf{v}_s$ and $\mathbf{v}_n$. For small velocities $\mathbf{j}$

can be expanded in powers of $\mathbf{v}_s$ and $\mathbf{v}_n$. In the first approximation

$$\mathbf{j} = \rho_s \mathbf{v}_s + \rho_n \mathbf{v}_n. \quad (7,2)$$

The coefficients ρ_s and ρ_n are obviously those which we called the densities of the superfluid and normal «parts» of the liquid. Their sum equals the real density ρ of helium II:

$$\rho = \rho_s + \rho_n, \quad (7,3)$$

ρ_s and ρ_n are, of course, functions of temperature. Note that the current $\mathbf{j}$ (7,2) is at the same time the density of the momentum, *i. e.* the momentum of a unit volume of liquid. ρ and $\mathbf{j}$ must satisfy the continuity equation:

$$\frac{\partial \rho}{\partial t} + \operatorname{div} \mathbf{j} = 0. \quad (7,4)$$

We shall here write the equations applicable to a motion in which the viscosity of the «normal liquid» plays no part. Then the equation for the momentum conservation is written in the form:

$$\frac{\partial j_i}{\partial t} + \frac{\partial \Pi_{ik}}{\partial x_k} = 0 \quad (7,5)$$

(the summation is extended over the indexes which are twice repeated), where the tensor Π_{ik} of the momentum current equals

$$\Pi_{ik} = p\delta_{ik} + \rho_n v_i^{(n)} v_k^{(n)} + \rho_s v_i^{(s)} v_k^{(s)}, \quad (7,6)$$

(p being the pressure). To take into account the viscosity of the normal liquid we must add to Π_{ik} the terms expressed in the ordinary way through the coefficients of viscosity and the derivatives of the velocity $\mathbf{v}_n$ with respect to the coordinates.

Further, the equation for the conservation of entropy takes the form:

$$\frac{\partial S\rho}{\partial t} + \operatorname{div}(\rho S \mathbf{v}_n) = 0 \quad (7,7)$$

(S is the entropy per unit mass of helium II). The «entropy current» equals $\rho S \mathbf{v}_n$, as the entropy is only transferred by the normal part of the liquid. If the viscosity of the normal part is taken into account supplementary terms must be added to the right-hand side of (7,7) expressing the increase of the entropy owing to the irreversibility of the processes.

Finally, the last equation of the complete set of hydrodynamic equations we get equalizing the accelleration $d\mathbf{v}_s/dt$ to the force acting on a unit of the «superfluid» mass. To determine this force imagine that the unit mass of liquid is displaced from the point 1 to the point 2 in such a way that the distribution of phonons and rotons is not changed. In other words, one might say that during the transfer only the «superfluid liquid» is displaced and the distribution of the normal liquid remains unchanged. The energy E of the liquid changes during such a transfer by

$$\left(\frac{\partial E}{\partial M}\right)_1 - \left(\frac{\partial E}{\partial M}\right)_2$$

(M being the mass of the liquid). Derivatives must be taken here at constant entropy (because the entropy is connected only with the normal liquid) and at a constant momentum of the motion of the normal mass of the liquid relative to the superfluid*; besides this the volume of the liquid is considered as a constant.

From the expression obtained for the change of energy it is seen that the quantity $\partial E/\partial M$ can be regarded as a «potential energy» of the superfluid liquid, so that the force acting upon it is

$$\operatorname{grad} \frac{\partial E}{\partial M}.$$

To calculate the derivative $\partial E/\partial M$ we notice that the derivative of the energy at constant entropy and volume is equal to the derivative of the thermodynamic potential at constant pressure and temperature. The thermodynamic potential $M\Phi$ of the liquid (Φ is the potential per unit mass) can be written in the form of the sum of the thermodynamic potential

* The motion of the superfluid liquid may be considered as external conditions in which the phonons and rotons move. Therefore, the «Lagrange function» for the motion of the normal liquid does not simply depend on its velocity $\mathbf{v}_n$, but on the difference of the velocities $\mathbf{v}_n - \mathbf{v}_s$. The conserved momentum is, therefore, a derivative of the Lagrange function with respect to $\mathbf{v}_n - \mathbf{v}_s$, *i. e.* the momentum of the relative motion.

$M\Phi_0(p, T)$ of the stationary liquid and the kinetic energy $\mathbf{P}^2/2M_n$ of the relative motion of the superfluid and normal «parts»:

$$M\Phi = M\Phi_0(p,T) + \frac{\mathbf{P}^2}{2M_n},$$

$\mathbf{P}$ is here the momentum of the motion of the normal mass M_n relative to the superfluid. By differentiating $M\Phi$ with respect to M at constant p, T and $\mathbf{P}$ and remembering that the normal mass M_n is proportional (at a given p and T) to the total mass M, we get

$$\Phi_0 - \frac{\mathbf{P}^2}{2M_n M}.$$

If we insert $\mathbf{P} = M_n(\mathbf{v}_n - \mathbf{v}_s)$ and put the ratio of the densities in place of the ratio of the masses we finally find, for the derivative $(\partial E/\partial M)_{S,V,\mathbf{P}}$ the expression

$$\Phi_0 - \frac{\rho_n}{2(\rho_n + \rho_s)}(\mathbf{v}_n - \mathbf{v}_s)^2.$$

It follows that the hydrodynamic equation for which we were looking is of the form:

$$\frac{d\mathbf{v}_s}{dt} = \frac{\partial \mathbf{v}_s}{\partial t} + (\mathbf{v}_s \nabla)\mathbf{v}_s = -\operatorname{grad}\left\{\Phi - \frac{\rho_n(\mathbf{v}_n - \mathbf{v}_s)^2}{2(\rho_n + \rho_s)^2}\right\}$$

(the index of Φ_0 is left out). It can be written differently if we notice (7,1), then

$$(\mathbf{v}_s \nabla)\mathbf{v}_s = \operatorname{grad}\frac{v_s^2}{2}$$

In this way

$$\frac{\partial \mathbf{v}_s'}{\partial t} = -\operatorname{grad}\left\{\Phi + \frac{v_s^2}{2} - \frac{\rho_n(\mathbf{v}_n - \mathbf{v}_s)^2}{2(\rho_n + \rho_s)}\right\}. \quad (7,8)$$

The equations (7,1)—(7,8) are a complete set of hydrodynamic equations for helium II.

For a stationary flow the left-hand side of (7,8) is zero; hence:

$$\frac{v_s^2}{2} - \frac{\rho_n(\mathbf{v}_n - \mathbf{v}_s)^2}{2(\rho_n + \rho_s)} + \Phi = \text{const.} \quad (7,9)$$

This equation together with the next one (7,10) plays here the rôle of the Bernoulli equation.

Consider now the motions at which liquid may be considered incompressible. If we take the densities ρ_n and ρ_s and entropy S as constants we find from (7,4) and (7,7) that

$$\operatorname{div}\mathbf{v}_s = 0, \quad \operatorname{div}\mathbf{v}_n = 0.$$

Now, for a stationary motion we have in (7,5) $\partial \mathbf{j}/\partial t = 0$; by using

$$\frac{\partial v_k^{(s)}}{\partial x_k} = \operatorname{div}\mathbf{v}_s = 0, \quad \frac{\partial v_k^{(n)}}{\partial x_k} = 0$$

we can rewrite (7,5) in the form

$$\nabla p + \rho_n(\mathbf{v}_n \nabla)\mathbf{v}_n + \rho_s(\mathbf{v}_s \nabla)\mathbf{v}_s = 0.$$

Remembering that $\operatorname{rot}\mathbf{v}_s = 0$ we can write this equation as

$$\nabla\left(p + \rho_n\frac{v_n^2}{2} + \rho_s\frac{v_s^2}{2}\right) = \rho_n[\mathbf{v}_n \operatorname{rot}\mathbf{v}_n].$$

We project this equation on the line of the current of the normal motion, *i. e.* on the direction of $\mathbf{v}_n$. Then on the right-hand side we get zero, so that

$$p + \rho_n\frac{v_n^2}{2} + \rho_s\frac{v_s^2}{2} = \text{const.} \quad (7,10)$$

It must be emphasized that the expression (7,10) is constant for a stationary flow only along each of the lines of current of the normal motion; and the expression (7,9) is constant over the whole volume of the liquid.

If the temperature and pressure change little over the volume of the liquid Φ can be expanded in powers of $\Delta T = T - T_0$, $\Delta p = p - p_0$; T_0, p_0 being the temperature and pressure at a certain point in the liquid:

$$\Phi = \Phi(p_0, T_0) - S\Delta T + \frac{\Delta p}{\rho}.$$

By inserting this into (7,9) we get

$$-\frac{v_s^2}{2} + \frac{\rho_n(\mathbf{v}_s - \mathbf{v}_n)^2}{2\rho} + S\Delta T - \frac{\Delta p}{\rho} = \text{const.}$$

By combining this equation with equation (7,10)

$$\Delta p + \frac{\rho_n v_n^2}{2} + \frac{\rho_s v_s^2}{2} = \text{const}$$

we get

$$\Delta T + \frac{\rho_n}{\rho S}\mathbf{v}_n(\mathbf{v}_n - \mathbf{v}_s) = \text{const.} \quad (7,11)$$

This relation, like (7,10), is valid along the current lines of the normal motion.

8. Propagation of sound in helium II

The equations obtained can be applied to the propagation of sound in helium II. The velocity of the motion in sound waves is as usual supposed to be small and the density, pressure and entropy are almost equal to their constant equilibrium values. The terms in (7,6) and (7,8) which are quadratic with respect to the velocities can be neglected, and in (7,7) we can take the entropy ρS in the term $\operatorname{div}(\mathbf{v}_n S\rho)$ out of the sign of div as this term already contains the small quantity $\mathbf{v}_n$. In this way the system of hydrodynamic equations for sound waves acquires the form

$$\frac{\partial\rho}{\partial t}+\operatorname{div}\mathbf{j}=0; \qquad (8,1)$$

$$\frac{\partial\rho S}{\partial t}+\rho S\operatorname{div}\mathbf{v}_n=0; \qquad (8,2)$$

$$\frac{\partial\mathbf{j}}{\partial t}+\nabla p=0; \qquad (8,3)$$

$$\frac{\partial\mathbf{v}_s}{\partial t}+\nabla\Phi=0. \qquad (8,4)$$

By differentiating (8,1) with respect to time and inserting (8,3) we get:

$$\frac{\partial^2\rho}{\partial t^2}=\Delta p. \qquad (8,5)$$

Further, we have

$$\frac{\partial S}{\partial t}=\frac{1}{\rho}\frac{\partial\rho S}{\partial t}-\frac{S}{\rho}\frac{\partial\rho}{\partial t}=-S\operatorname{div}\mathbf{v}_n+\frac{S}{\rho}\operatorname{div}\mathbf{j},$$

or

$$\frac{\partial S}{\partial t}=\frac{S\rho_s}{\rho}\operatorname{div}(\mathbf{v}_s-\mathbf{v}_n). \qquad (8,6)$$

For the thermodynamic potential the relation

$$d\Phi=-SdT+Vdp=-SdT+\frac{1}{\rho}dp$$

holds (V being the specific volume). Hence we have

$$\nabla p=S\rho\nabla T+\rho\nabla\Phi,$$

or by introducing ∇p from (8,3) and $\nabla\Phi$ from (8,4)

$$\rho_n\frac{\partial}{\partial t}(\mathbf{v}_n-\mathbf{v}_s)+\rho S\nabla T=0. \qquad (8,7)$$

Differentiating (8,6) with respect to time and introducing (8,7) we find:

$$\frac{\partial^2 S}{\partial t^2}=\frac{S^2\rho_s}{\rho_n}\Delta T. \qquad (8,8)$$

Two equations (8,5) and (8,8) determine the propagation of sound in helium II. It is already seen from the fact that there are two equations that there must be two velocities of sound in heluim II.

Write S, ρ, p, T in the form $S=S_0+S'$, $\rho=\rho_0+\rho'$, etc. where the quantities with a dash represent the small changes of the corresponding quantities stipulated by the sound wave and the quantities with index zero are their constant equilibrium values. Then we can write:

$$\rho'=\frac{\partial\rho}{\partial T}T'+\frac{\partial\rho}{\partial p}p', \quad S'=\frac{\partial S}{\partial T}T'+\frac{\partial S}{\partial p}p',$$

and equations (8,5) and (8,8) take the form

$$\frac{\partial\rho}{\partial p}\frac{\partial^2p'}{\partial t^2}-\Delta p'+\frac{\partial\rho}{\partial T}\frac{\partial^2T'}{\partial t^2}=0,$$

$$\frac{\partial S}{\partial p}\frac{\partial^2p'}{\partial t^2}+\frac{\partial S}{\partial T}\frac{\partial^2T'}{\partial t^2}-\frac{S^2\rho_s}{\rho_n}\Delta T'=0.$$

We look for a solution of these equations in the form of a plane wave in which p' and T' are proportional to a factor $e^{i\omega(t-x/u)}$ (u being the velocity of sound) and then for the conditions of solubility we get the equation:

$$\begin{vmatrix} u^2\frac{\partial\rho}{\partial p}-1 & u^2\frac{\partial\rho}{\partial T} \\ u^2\frac{\partial S}{\partial p} & u^2\frac{\partial S}{\partial T}-S^2\frac{\rho_s}{\rho_n} \end{vmatrix}=0,$$

or

$$u^4\frac{\partial(\rho,S)}{\partial(p,T)}-u^2\left(\frac{\partial S}{\partial T}+S^2\frac{\rho_s}{\rho_n}\frac{\partial\rho}{\partial p}\right)+S^2\frac{\rho_s}{\rho_n}=0$$

[where $\partial(\rho,S)/\partial(p,T)$ denotes the Jacobian of the transformation from ρ, S to p, T]. By means of a simple transformation with the use of the thermodynamic relations this equation can be put in the form

$$u^4 - u^2\left[\left(\frac{\partial p}{\partial \rho}\right)_S + \frac{TS^2}{C_v}\frac{\rho_s}{\rho_n}\right] + \frac{S^2\rho_s T\left(\frac{\partial p}{\partial \rho}\right)_T}{\rho_n C_v} = 0 \qquad (8,9)$$

(C_v being the heat capacity of a unit mass of helium II). This quadratic equation determines two velocities of sound in helium II.

If $\rho_s = 0$, *i. e.* at the λ-point, one of the roots of the equation (8,9) becomes zero and we get, as we ought, only one ordinary velocity of sound

$$u = \sqrt{\left(\frac{\partial p}{\partial \rho}\right)_S}.$$

Practically, for all temperatures the heat capacities C_p and C_v are close to each other. According to the known thermodynamic formula in these conditions the isothermic and adiabatic compressibilities are also close to each other, *i. e.*

$$\left(\frac{\partial p}{\partial \rho}\right)_S \approx \left(\frac{\partial p}{\partial \rho}\right)_T.$$

If we denote the common value of $\left(\frac{\partial p}{\partial \rho}\right)_T$ and $\left(\frac{\partial p}{\partial \rho}\right)_S$ as $\frac{\partial p}{\partial \rho}$ and the common value of C_p and C_v simply as C we get from the equation (8,9) two velocities of sound u_1 and u_2 in the form

$$u_1^2 = \frac{\partial p}{\partial \rho}, \quad u_2 = \frac{TS^2}{C}\frac{\rho_s}{\rho_n}. \qquad (8,10)$$

In this way one of the velocities (u_1) is almost constant and the other (u_2) strongly depends on the temperature becoming zero at the λ-point. At a temperature 1.33° K we get a value of about 25 m/sec. for u_2. At extremely low temperatures, when $\rho_n^{(ph)} \gg \rho_n^{(r)}$ one gets

$$u_2 = \frac{c}{\sqrt{3}} \qquad (8,11)$$

In this way as the temperature tends to zero the velocity of sound tends to constant limits $u_1 = c$, $u_2 = c/\sqrt{3}$.

9. The problem of superconductivity

The phenomenon of superconductivity is in many ways akin to the phenomenon of superfluidity. Superconductivity can also be explained by the supposition of the presence of an energy gap in the spectrum of the «electron liquid» in a metal. However, the character of this gap differs from that which is found in liquid helium. The raising of the question itself concerning the division of the motions of a liquid into potential and vortex is senseless when applied to the electron liquid in a metal.

The idea of a connection between superconductivity and the presence of a gap in the energy spectrum has been advanced several times [cf., for example, ([18])]. However, there has been no clear proof given that the presence of this gap really leads to the phenomenon of superconductivity.

In the electron liquid in a metal one can distinguish between the motions of a pure inner character and the motions in which the liquid moves as a whole, in a macroscopic manner. The latter correspond to the presence of a total current in the metal. The states with such a «macroscopic» motion begin directly from the normal unexcited state*. To explain superconductivity one must suppose that there is an energy interval between the normal state of the electron liquid and the beginning of the continuous spectrum of the states with an inner motion.

The value of this interval must be very small,—in a temperature scale of the order of temperatures of the superconductive transition. What causes this gap is not, up to the present time, clear.

Let Ψ_0 be the wave function of the normal state of the electron liquid in which it is found at absolute zero. We shall consider a certain excited state which is close to the normal one. According to what was said this state is the state of the motion of the electron liquid as a whole. Corresponding to this the wave function Ψ of this state should have the form

$$\Psi = \Psi_0 e^{\frac{i}{\hbar}\mathbf{p}\sum_a \mathbf{r}_a},$$

where $\mathbf{p}$ is the momentum of the macroscopic motion per electron and the sum-

* In this we have the essential distinction between superconductors and insulators in which there is also an energy gap, however, no states corresponding to a presence of a total current exist close to the normal state.

mation is extended over the radius-vectors of all the electrons.

Actually, however, one must keep in mind that a motion of the electron liquid as a whole must not necessarily have the same velocity in its whole volume. Therefore, in the wave function Ψ the exponential factor actually has the form $\exp\left(\frac{i}{\hbar}\sum\chi_a\right)$ and not $\exp\left(\frac{i}{\hbar}\mathbf{p}\sum\mathbf{r}_a\right)$; χ is a certain function of the coordinates of one of the electrons about which one can only say that it is nearly linear; different terms in the sum $\sum\chi_a$ differ in that χ is taken as a function of the coordinates $\mathbf{r}_a$ of different electrons

$$\Psi = \Psi_0 e^{\frac{i}{\hbar}\sum_a \chi_a}. \quad (9,1)$$

We determine the density $\mathbf{j}$ of the current in the electron liquid in the state Ψ. To do this we have to integrate the expression

$$\frac{\hbar e}{2im}(\Psi^*\nabla_a\Psi - \Psi\nabla_a\Psi^*) - \frac{e^2}{mc}\mathbf{A}\Psi\Psi^*$$

(m being an «effective mass» of the electron, $\mathbf{A}$ — the vector potential of the magnetic field, e — the electron charge, c — the velocity of light) over the coordinates of all the electrons except the ath and then a summation must be made over all a (*i. e.* over all the electrons), taking the value of each term at the same point of the liquid. For $\Psi = \Psi_0$ and $\mathbf{A} = 0$, *i. e.* for an electron liquid at rest, the current $\mathbf{j}$ obviously equals zero. The result is

$$\mathbf{j} = N\frac{e}{m}\,\mathrm{grad}\,\chi \cdot \int |\Psi_0|^2 d\tau' - N\frac{e}{mc}\mathbf{A}\int |\Psi_0|^2 d\tau'$$

($d\tau'$ is the product of the differentials of coordinates of all the electrons except of one of them; N is the total number of electrons). But the integral

$$\int |\Psi_0|^2 d\tau' = \int |\Psi|^2 d\tau'$$

is the probability density of finding electron at a given point of space. Therefore, the product $N\int|\Psi|^2 d\tau'$ is the number n of electrons in a unit volume of the electron liquid. In this way

$$\mathbf{j} = \frac{ne}{m}\,\mathrm{grad}\,\chi - \frac{e^2}{mc}n\mathbf{A}. \quad (9,2)$$

Excluding χ we find

$$\mathrm{rot}\,\frac{mc\mathbf{j}}{e^2 n} = -\mathrm{rot}\,\mathbf{A} = -\mathbf{B} \quad (9,3)$$

($\mathbf{B} = \mathrm{rot}\,\mathbf{A}$ is the magnetic induction). This equation may be regarded as a supplementary condition which must be satisfied in the possible states of the macroscopic motion of the electron liquid. It is to be noted that this relation is a generalization of the equation $\mathrm{rot}\,\mathbf{v}_s = 0$ which is found for superfluidity and goes over into it if the magnetic field can be neglected.

If the metal is homogeneous the electron density can be considered constant and the equation (9,3) can also be written as

$$\mathrm{rot}\,\mathbf{j} = -\frac{ne^2}{mc}\mathbf{B}. \quad (9,4)$$

From this equation [first put forward by R. Becker, G. Heller, and F. Sauter [14] and first written in the form (9,4) by London [15]] the magnetic (Meissner effect) and the electric (lack of resistance) properties of superconductors follow. It is known also that (9,4) is also supported quantitatively by the experiments on the superconductivity of helium films [16].

Similarly to the conclusions in § 5 one can come to the conclusion that at non-zero temperatures the equation (9,4) still holds with the only difference that en now equals not the total charge density of the electron liquid but only a certain portion of it. In this way one can in a certain sense speak of the division of the whole charge density into «superconducting» and «normal» parts, but not forgetting all the reservations mentioned for the similar division of the density of the mass of liquid helium. In particular, this does not mean that there are «superconducting» and «normal» electrons in metals; simply

the motion of the electron liquid in the metal can be described as a combination of two simultaneous motions each of which is connected with a certain effective charge. The superconductivity vanishes at that temperature at which the «superconducting» charge becomes zero

It must be noted, however, that in the equation (9,4) the charge *en* enters in a combination *ne/m* with the mass *m*. Therefore, only that ratio has a physical sense and not each one of the values of *en* and *m* separately.

As in helium II we come to the conclusion that the superconducting current must not transfer heat. This is supported by the well known fact that the thermoelectric phenomena are absent in superconductors ([17]).

In conclusion I wish to express my thanks to P. L. Kapitza for kindly informing me of his experimental results before publication.

**Institute for Physical Problems,
Academy of Sciences of the USSR.
Moscow.**

REFERENCES

1. **P. L. Kapitza, Nature, 141, 74 (1932).**
2. **L. Tisza, Nature, 141, 913 (1938).**
3. **A Bijl, Physica, 7, 869 (1940).**
4. **W. H. Keesom, Leiden Comm., Suppl., 80b (1936)**
5. **W. H. Keesom a. A. P. Keesom, Physica, 2, 557 (1935).**
6. **B. Bleaney, F. Simon, Trans. Faraday Soc., 35, 1205 (1939).**
7. **L. Landau a. E. Lifshitz, Statistical Physics, Oxford, 1938.**
8. **W. Keesom, G. McWood, Physica, 5, 257 (1938).**
9. **P. L. Kapitza, Journ. of Phys., this issue, p. 59.**
10. **H. London, Proc. Roy. Soc. (A), 171, 484 (1939).**
11. **W. H. Keesom a. Miss Keesom, Physica, 3, 359 (1936).**
12. **P. L. Kapitza, Journ. of Phys., 4, 181 (1941).**
13. **H. Welker, ZS. Physik, 114, 535 (1939).**
14. **R. Becker, G. Heller, F. Sauter, ZS. Physik, 85, 772 (1933).**
15. **F. a. H. London, Proc. Roy. Soc. (A), 149, 71 (1935).**
16. **E. T. S. Appleyard, J. T. Bristow, Proc. Roy. Soc. (A), 172, 530 (1939); A. Shalnikov, Nature, 142, 74 (1938).**
17. **W. H. Keesom, G. J. Matthijs, Physica, 5, 437 (1938).**

LETTERS TO THE EDITOR

ON THE THEORY OF SUPERFLUIDITY OF HELIUM II

By L. Landau

Institute for Physical Problems, Academy of Sciences of the USSR

(Received November 15, 1946)

The velocity of the "second sound" in helium II has been measured by V. Peshkov [1] with a great precision. His results give an opportunity to perform a quantitative comparison of the theory developed by the author [2] with the experiment. Such a comparison gives full support to the general picture given by the theory, but at the same time it reveals a noticeable discrepancy between the calculated and observed values of the velocity (*e. g.* 25 m/sec. calculated and 19 m/sec. observed at the temperature of 1.6°K). Although this discrepancy is not very large, it is too large to be attributed to the inaccuracy of the experimental data on the thermodynamic quantities of helium II.

For calculating the velocity of the second sound the formulae were used for the thermodynamic quantities (entropy, specific heat), derived in [2] under the assumption of the energy spectrum of the liquid to consist of two branches — the phonon and roton ones. The direction of the observed discrepancy indicates in what way these assumptions must be altered. Using the experimental data, one can formally compute the roton mass μ according to formulae

$$\rho_n = N\mu, \quad F_r = -NkT. \tag{1}$$

Here N is the number of rotons per unit volume, F_r—the "roton part" of the free energy (*i. e.* the free energy without the vibrational part), ρ_n—the density of the "normal part" of the liquid (the phonon part in ρ_n is negligible as compared with the roton part). The mass μ calculated in this way appears to be approximately inversely proportional to the temperature (in temperature interval 1.3—1.7°K), instead of being constant. It is, however, to be noted, that although the very fact of the variation of μ is apparent, the quantitative law of its variation can be established only in a very approximate way (owing mainly to the scarcity of experimental data on the specific heat of helium II).

If one does not make the assumption $\varepsilon = \Delta + p^2/2\mu$ for the dependence of the energy ε of a roton on its momentum p, but considers the general dependence $\varepsilon(p)$, then the calculation according to the general formulae derived in [2] shows, that in the formula $\rho_n = N\mu$ the quantity $\overline{p^2}/3kT$ enters instead of μ ($\overline{p^2}$ is the mean square of the momentum). If this quantity is inversely proportional to the temperature, then $\overline{p^2}$=const., *i. e.* the values of the roton momenta lie mainly in the neighbourhood of a certain p_0. At the first glance this fact appears to be very strange, but it can be explained in a natural manner by assuming, that the energy spectrum of helium II is of the type shown in Fig. 1. For small momenta p of an

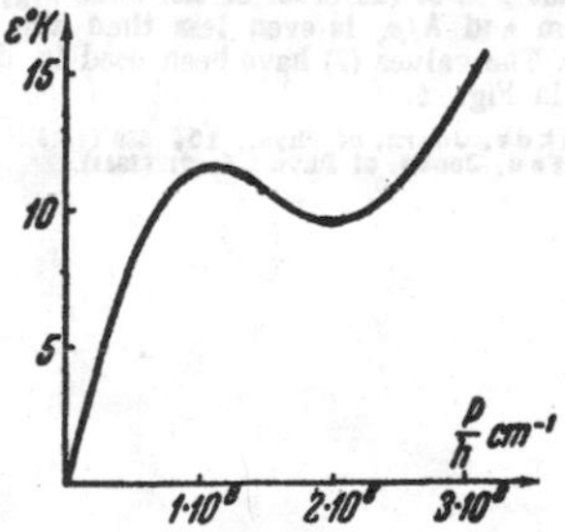

Fig. 1

elementary excitation its energy ε increases linearly (phonons), then reaches a maximum, begins to decrease and at a certain value $p = p_0$ the function $\varepsilon(p)$ has a minimum. In the neighbourhood of the latter we can write

$$\varepsilon = \Delta + \frac{(p-p_0)^2}{2\mu}. \tag{2}$$

μ being a constant. With such a spectrum it is of course impossible to speak strictly of rotons and phonons as of qualitatively different types of elementary excitations. It would be more correct to speak simply of the long wave (small p) and short wave (p in the neighbourhood of p_0) excitations. It is to be stressed, that all the conclusions concerning the superfluidity and the entire macroscopical hydrodynamics of helium II, developed in [2], maintain their validity also with the spectrum proposed here.

Only the formulae for the thermodynamic quantities must be changed. Instead of formulae (3), (4)—(7) in [2] we have for the "roton" parts of the free energy, entropy, specific heat (per unit mass) and the density of the "normal liquid":

$$F_r = -\frac{2\mu^{1/2}(kT)^{3/2}p_0^2}{(2\pi)^{3/2}\rho\hbar^3}e^{-\Delta/kT}; \qquad (3)$$

$$S_r = \frac{2(k\mu)^{1/2}p_0^2\Delta}{(2\pi)^{3/2}\rho T^{1/2}\hbar^3}\left(1+\frac{3kT}{2\Delta}\right)e^{-\Delta/kT}; \qquad (4)$$

$$C_r = \frac{2\mu^{1/2}p_0^2\Delta^2}{(2\pi)^{3/2}\rho k^{1/2}T^{3/2}\hbar^3}\times$$
$$\times\left[1+\frac{kT}{\Delta}+\frac{3}{4}\left(\frac{kT}{\Delta}\right)^2\right]e^{-\Delta/kT}; \qquad (5)$$

$$\frac{(\rho_n)_r}{\rho} = \frac{2\mu^{1/2}p_0^4}{3(2\pi)^{3/2}\rho(kT)^{1/2}\hbar^3}e^{-\Delta/kT}. \qquad (6)$$

In such a form the theory contains three constants: Δ, p_0 and μ. It is, therefore, difficult to check it on the basis of the experimental data which are now available. For the values of Δ, p_0 and μ one gets:

$$\frac{\Delta}{k} = 9.6°, \quad \frac{p_0}{\hbar} = 1.95\cdot 10^8\ \text{cm}^{-1}, \quad \mu = 0.77\, m_{He}. \qquad (7)$$

Note that μ is of the order of the mass m_{He} of the helium atom and $\hbar/p_0$ is even less than the atomic dimensions. The values (7) have been used in drawing the curve in Fig. 1.

[1] V. Peshkov, Journ. of Phys., 10, 389 (1946).
[2] L. Landau, Journ. of Phys., 5, 71 (1941).

Printed in the United States
by Baker & Taylor Publisher Services

Printed in the United States
by Baker & Taylor Publisher Services